What is Environmental Politics?

CAULTfield

What is Environmental Politics?

ELIZABETH R. DESOMBRE

polity

First published in 2020 by Polity Press

Polity Press
65 Bridge Street
Cambridge CB2 1UR, UK

Polity Press
101 Station Landing
Suite 300
Medford, MA 02155, USA

ISBN-13: 978-1-5095-3413-5
ISBN-13: 978-1-5095-3414-2 (pb)

A catalogue record for this book is available from the British Library.

Library of Congress Cataloging-in-Publication Data

Names: DeSombre, Elizabeth R., author.
Title: What is environmental politics? / Elizabeth R. DeSombre.
Description: Cambridge, UK ; Medford, MA : Polity, 2020. | Series: What is politics | Includes bibliographical references and index. | Summary: "Scientific knowledge and technology alone cannot address environmental problems; they also involve difficult political choices and trade-offs both locally and globally. This concise introductory text explores the different ways in which society attempts to deal with the political decisions needed to prevent or recover from environmental damage"-- Provided by publisher.
Identifiers: LCCN 2019029829 (print) | LCCN 2019029830 (ebook) | ISBN 9781509534135 (hardback) | ISBN 9781509534142 (paperback) | ISBN 9781509534159 (epub)
Subjects: LCSH: Environmentalism--Political aspects. | Environmental protection--Political aspects. | Environmental policy.
Classification: LCC JA75.8 .D488 2020 (print) | LCC JA75.8 (ebook) | DDC 363.7--dc23
LC record available at https://lccn.loc.gov/2019029829
LC ebook record available at https://lccn.loc.gov/2019029830

Typeset in 11 on 14 pt Sabon
by Servis Filmsetting Ltd, Stockport, Cheshire
Printed and bound in the UK by TJ International Limited

For further information on Polity, visit our website: politybooks.com

Contents

Acknowledgements

Writing a book is in some ways a solitary activity, but it wouldn't be possible without a community. I am grateful to be embedded in several communities without which this book wouldn't have been possible. The first is the Environmental Studies Department at Wellesley College, a collaborative community of faculty, students, and staff in which most of the ideas presented here first emerged, both inside and outside of the classroom. I know from experience that not all academic units are as supportive and friendly as my current one is, and I'm beyond grateful for fascinating colleagues and smart students. (And for smart colleagues and fascinating students, too.)

I also benefit from being a part of the Environmental Studies Section of the International Studies Association, another set of scholars who both challenge and encourage each other. A related community is the Teaching Global Environmental Politics (GEP-ED) listserv, where queries about things such as environmental successes or publisher marketing surveys are quickly and helpfully answered, and I learn from listening to the various debates that others bring. I am not a natural networker, and the fact that people around the world who study and teach the kinds of things I do are people I want to hang out with makes our collective engagement much less instrumental and much more enjoyable.

Thanks are also due to several individuals: thanks to Louise Knight at Polity Press for pitching the idea of this book to me.

I wouldn't have thought to propose it, but once she asked I realized I had some strong opinions on what environmental politics is, and isn't, and I thought that it would be fun to communicate them. It has been. Bridget Peak provided excellent research assistance and also feedback on an earlier draft of the book. Sammy Barkin reads and comments on everything I write, and all my ideas and information are better because of his feedback. Lynda Warwick is the perfect non-academic audience for my thoughts and one of my favorite people. She helps make sure I can communicate academic work to non-specialists and also just generally helps me cope with life. As does Zoë, whose main contribution to this book involved competing with my computer for lap space and ensuring that I do my big picture thinking on lots of long walks.

CHAPTER 1

Defining Environmental Politics

Why is it so difficult to prevent, or fix, problems of pollution?
Why do we continually harvest or extract natural resources at
unsustainable rates, even though these activities cause known
harm to both people and ecosystems? Why do environmental
problems frequently harm groups of people who are already
the most vulnerable and least powerful?

Addressing environmental issues isn't primarily an issue
of science or information. Despite uncertainty about some
details, we frequently understand the basic processes by which
environmental problems happen, at least once we discover
them. Responding to – or even creating – environmental problems requires political decisions: choices by governments (and
others) about how to allocate resources and prioritize tradeoffs
among different social values. These political decisions involve
advocacy by groups with various priorities and different levels
of political influence. This political process, which can be difficult enough for many social issues, faces particular challenges
when addressing the environment.

Environmental politics is the study of how societies make
decisions about resource use, pollution, or economic activities that influence the condition of the environment. These
processes frequently, but not always, take place within government institutions. They involve or affect people from
many economic and social sectors. Preventing or addressing
environmental problems requires accurately diagnosing the

characteristics of these problems and understanding the political processes for making rules about the activities that cause them.

What is the Environment?

For a term that is so frequently used, it can be surprisingly difficult to pin down what "the environment" means. Some people use it to refer to nature and the various amenities that nature provides: plants, animals, soils, air, water, even sunlight. But the concept is broader than that and changes as we imagine different human activities and learn more about global systems, from the smallest to the largest scale. We now understand the importance of microbes and of the stratospheric ozone layer in protecting life on earth. At the same time, human constructs such as buildings and cities, or practices such as agriculture or industrialization, form part of what we are thinking of when we talk about the environment.

What we are generally discussing when we look at the environment is how human activity influences it, and is influenced by it. One useful framing is to think of the environment in this context as being partly about pollution, which occurs when contaminants or other unwanted substances are introduced (usually unintentionally) into air, water, or land as a byproduct of other activity. It is also partly about the use of resources, things taken out of their environmental context for human use. These resources may be renewable or non-renewable.

Pollution is pervasive. It includes everything from waste dumped intentionally into the ocean to fumes from factory smokestacks to agricultural runoff into rivers. We can even speak of light pollution or noise pollution to refer to unwanted intrusions into our surroundings that affect the wellbeing of people or other species.

Some pollution is obvious. We can see or smell it. We may know its causes and be able to feel its effects. Other types of pollution, however, are less apparent. Water that is contaminated by lead or other heavy metals may be toxic to people and animals, but it may look or taste no different than clean water. Greenhouse gases – carbon dioxide and other emissions that contribute to global climate change – are substances that, on their own, aren't necessarily harmful to breathe. Collectively they nevertheless contribute to the overall warming of the planet and the many related effects such as drought and sea-level rise. The chlorofluorocarbons (CFCs) that contributed to the destruction of the ozone layer were non-toxic and stable and invisible in the air. Pollution, in other words, is defined by the harm it causes, but we may not always be aware of that harm when we are creating or experiencing it.

Natural resources also form a part of what we think of as the environment. Overuse of natural resources is an important category of environmental problems. Some natural resources are what we call "non-renewable." On human time-scales the earth will not make any more of these materials than currently exist. Non-renewable resources include things such as fossil fuels – the coal, oil, and gas that are so central to our energy use – as well as minerals such as copper and tin, or even the rare earth elements (for example, neodymium) that are important in our cell phones and other technology.

Other types of natural resources – such as fish or forests – are what we call "renewable." You'd think that use of renewable resources would be less problematic than use of non-renewable resources. After all, if you leave enough fish or trees around, they make more fish and trees. But it turns out that we often do extract these far beyond their ability to be renewed.

There is more to the environment than decisions about endangered species or factory pollution; these include factors such as urbanization or agriculture. At this point in ecological history, humans have shaped almost every aspect of the earth's processes, so much so that geologists and others argue that we've entered a new geologic age: the Anthropocene. This era is named to reflect the pervasive influence of humans on global processes. Those effects are all relevant to a consideration of what the environment is.

Also relevant are questions of who the environment is *for*. When we're concerned about protecting the environment, do we care only about ensuring its best uses for the wellbeing of people, or do we care about species or resources for their own sake? From the point of view of environmental politics, the focus is most frequently on people, because people are the decision makers, and because politicians are elected or otherwise maintained in power by the support of the people they govern. But species and ecosystems may matter in their own right and not just for their instrumental value to human beings, and some people do focus on the wellbeing of the environment or of certain species for the sake of ecosystems or those species and engage in environmental politics to protect them.

Even if we are concerned about the condition of the environment primarily for the sake of people, we need to ask which people bear the brunt of environmental problems. The concept of environmental injustice has been framed to call attention to the fact that some of the most vulnerable populations are the most likely to suffer the worst consequences of environmental problems and have the least voice in the creation of policies to address the environment. An understanding of characteristics of both politics and the environment can help us appreciate why it is that environmental problems are often concentrated in communities with the least political power.

What's Special about the Environment?

Environmental issues have some unusual characteristics that have implications for our political efforts to address them. No one intends to create environmental problems, and the effects of these problems are often felt far away both in space and in time from where they are caused. Most environmental problems come from the combined actions of many people, which means that each person's contribution to the problem feels small, and addressing the problem requires changing the behavior of a lot of people. So no one person can fix or prevent environmental problems alone, and, at the same time, those who don't contribute to addressing these problems can undermine the ability of others to do so. While the environment isn't the only issue area that has each of these characteristics, taken together they help explain why it's fairly easy to create environmental problems and difficult to take successful action to prevent or fix them. These characteristics interact with the political structures and actors discussed in this book to help explain environmental action or inaction across different political jurisdictions.

Externalities

Environmental problems are externalities: unintended, and unpriced, consequences of other things people are trying to do. Outside of bad superhero movies, no one wakes up in the morning and decides to create smog or deplete an entire species. No one intends to cause, or even contribute to, global climate change.

Likewise, no one plans to contribute to a huge patch of plastic in the middle of the Pacific Ocean; people are simply seeking a way to carry groceries home or drink iced tea. The plastic bags or cups that may be given out for free allow people to accomplish those goals but, eventually, create a set of harms the user did not intend and may not even be aware of. The pollution caused from generating the energy people use is not something anyone plans or even wants to create; we simply want to be able to heat our homes, transport our family members, or see at night.

Externalities are experienced primarily by people other than the ones who get the benefit of the activity that caused them. That means the people doing an activity rarely take its externalities into consideration. If every time someone drank coffee in a disposable cup another tree in her yard disappeared and her drinking water became more contaminated with chemical pollution, she would quickly decide not to use disposable cups. Instead, those effects are most likely felt by people far away, in both space and time, from the coffee drinker. She is probably unaware of them and doesn't directly experience any of the downsides of her cup use.

Externalities can be either positive or negative. You can create unintended consequences from your activities that are beneficial to others; they are externalities because they are not intended when you choose to undertake the activity, and they do not affect the cost of your actions. Someone who plants flowers for her own enjoyment may create positive externalities in the neighborhood; those who pass by may enjoy seeing or smelling the flowers, and the flowers may create beneficial habitat for butterflies or bees.

The flower example illustrates another concept: whether an externality is positive or negative depends on the perspective of those who experience it. The same flowers that give one neighbor pleasure may contribute to the allergies of a different neighbor. When we discuss the role of externalities in creating environmental problems, we are concerned primarily about negative externalities, so those are the ones that are discussed in this book.

The "unpriced" aspect of externalities has several important implications. First, it means that the person causing the
externality doesn't bear a cost for doing it. The sulfur dioxide
pollution from coal burned to create electricity doesn't factor
into the price paid by the electricity generator or the consumer.
That doesn't mean that there isn't a cost from that pollution.
People get asthma and generate costs for doctors and drugs;
employers lose money when their employees take sick days.
The acid rain that results causes buildings to disintegrate and
plants to die, among many other effects. Someone bears that
cost, but it is not the generator of the electricity or (primarily)
the people who use it.

In other words, it doesn't cost any more to do what you're
doing in a way that causes pollution than in a way that
doesn't. For that reason, most people don't even notice that
they're causing externalities (which is different from how they
might notice if they left a water faucet on and would therefore
get a much higher water bill). In fact, precisely because of the
unpriced nature of externalities, it would almost always cost
more to stop creating externalities than to continue to create
them, at least initially, for whoever is creating them.

People don't experience much, if any, of the harm from the
externalities to which they contribute; that's part of why they
are considered to be *external*. There are several reasons that
those who create them are unlikely to suffer from these externalities. The first is that there's usually a disconnect in time
and space from where an action takes place and where the
results are felt. The sulfur dioxide emissions from a coal-fired
power plant travel hundreds of miles in the air from where
they are emitted, so the people experiencing their effects are
rarely the same people using the electricity. (Even if the effects
are felt locally, they are felt by many people, regardless of how
much electricity each uses.) Other environmental issues take

a while to be felt. Ozone depletion was caused by chemicals (used in refrigeration and electronics production) that had to make their way a long distance into the stratosphere, and to accumulate in significant enough quantities, to make a difference that we would notice on earth. Some of those chemicals, like some of the substances that cause global climate change, can persist for hundreds of years or more in the atmosphere, causing environmental problems generations after they were initially released.

Some externalities are more removed than others from the activities that create them. Someone who fishes does not intend to cause the depletion of a fishery but does intend to take fish. Someone using nitrogen fertilizers in the Midwest of the United States is simply trying to grow crops more successfully, conceptually unrelated to the dead zones in the Gulf of Mexico that result when too much nitrogen or phosphorus in the water causes algal blooms that use the available oxygen and make areas of the ocean unable to support life. How closely connected an activity is to the externality it creates can influence the likelihood of causing it in several ways, which are discussed further below.

Figuring out how to prevent externalities is key to avoiding or addressing environmental problems. Because of the nature of environmental problems as externalities, finding ways to accomplish the same underlying goal without creating the externality is likely, at least initially, to cost more. That is because of the "unpriced" aspect of externalities. If there is no cost to putting sulfur dioxide into the air, then burning coal to produce electricity gets all the benefits of the electricity for those who produce or use the electricity without their having to pay for the downsides of that coal burning. Generating electricity in a way that doesn't create sulfur dioxide emissions requires either using a different, more costly, input in areas where coal may be cheap or installing "scrubbers" to take the sulfur dioxide out of the emissions; those scrubbers have a cost and also make the electricity generation less efficient. Producing electricity in a different way is therefore likely to be more costly; if that were not the case, the change would have likely already been made. (That's the upside of the unintended aspect of externalities: because you're not *trying* to create them, if it were possible to do the thing you're trying to do just as easily without creating them, you'd be happy to do it.)

That additional cost means that people or businesses are not likely to change their behavior to avoid creating externalities of their own accord. Policy can therefore play a central role in making that change happen; in many countries power plants are required by law to remove the sulfur dioxide from their emissions. But in the same way that people or businesses resist deciding on their own to stop creating externalities, they are likely to oppose policy action to prevent them from creating externalities. That's where politics comes in – the struggle among different people with differing opinions on what should be done. The question of what businesses or people should be required to do, or prohibited from doing, is a political decision.

The economist Ronald Coase argued that externalities can sometimes be addressed without policy intervention. His logic rests on the important observation that externalities are reciprocal. They connect at least two different actors: the one creating the externality and the one it affects. One of the examples he uses is the effect of a cattle ranch next to a farm; the straying cattle can trample the crops and cause damage to the farm. That damage is an externality; it is unintended and the rancher doesn't suffer any cost from the straying cattle unless someone creates a rule that requires the herder to compensate for, or prevent, the damage. As is obvious, there would be no damage to the crops without the straying cattle. But, as

Coase also points out, it is also true that "there would be no crop damage without the crops."[1] In other words, the rancher is affected by the presence of the farmer as well as the other way around.

This reciprocal relationship means that, in the same way that the farmer is negatively affected by the behavior of the rancher, the rancher would be negatively affected by having to change her behavior to avoid the damage her cows are doing to the crops. If there is value in ranching, to the rancher or to the community, then simply requiring the rancher to stop creating the externality may not be the best collective solution.

The most common way to address problems created from externalities is to regulate the action that is producing them. The rancher, in this instance, could be compelled to fence her property or be required to pay compensation for any crop damage. The reciprocal element of externalities creates opportunities to address these unintended consequences in ways other than governmental intervention, however. The farmers could work together to put up a fence to keep the cows out. If the cost of the fence was less than the cost of the damage from the cows, it might be worthwhile to the farmers to work together to put one up, especially if there were no existing rule that prevented the rancher from causing damage. The farmers could even pool their money to offer a certain amount to the rancher to persuade her not to buy another cow. For the farmers, again, this solution would be worthwhile if the amount of damage avoided would be greater than the cost they would have to pay. For the rancher, it would be worthwhile to take money in return for not getting another cow if the amount earned from the cow would be less than the amount the farmers offered. There might be a situation in which both parties are better off than either would be without that privately agreed solution.

Depending on your perspective, creating this type of solution might not appear "fair." It may not seem reasonable that a farmer who is being harmed by someone else's activity should have to be the one to take on the cost of preventing the problem. But what Coase is pointing out is that, in the absence of a political solution that regulates the rancher, the farmers still have some ability to improve their situation on their own. And, to the rancher, a newly created set of farms that affect her ranching operations also might not seem "fair."

"Coasian" solutions to the problem – addressing externalities without government action – do have some conditions that need to be met before they are likely to happen. First, everyone needs to have full information about the costs and the benefits of the externality and any potential solutions to it. Second, any agreement that the parties reach needs to be enforceable; if the farmers pay the rancher not to get another cow and the rancher gets one anyway (and there's no recourse), this type of solution will not be pursued in the future. Third, what Coase calls "transaction costs" – the difficulties and actual costs of pursuing the solution – need to be minimized.

Even if the conditions Coase outlines are not likely to be met in most cases of environmental externalities, there are some important implications of his argument. First, working to improve those conditions can be useful not only in their own right but because they can help communities be more willing to address environmental externalities on their own. Reducing transaction costs – perhaps by holding a neighborhood meeting and providing childcare – can make it easier for the sufferers of an externality to organize. Making information transparent is a good thing in its own right and can help the process of figuring out what the best solution collectively would be. Doing these things can also make communities less resistant to policy to address the externalities in the first place, because they better understand both the actual long-term costs and benefits of changing behavior and the underlying environmental problem.

Finally, implicit in the Coase Theorem is the idea that we don't necessarily want to aim for a situation in which no externalities are produced at all. That sounds counterintuitive: if externalities are negative, wouldn't we want them to be eliminated? But the activity that creates externalities frequently has value, and efforts to reduce the externalities produced will also reduce the amount of that activity. Electricity generation frequently causes environmental damage, but electricity is central to important endeavors. A manufacturing plant that produces pollution may be making life-saving medical equipment. Stopping electricity generation or manufacturing in order to prevent the externalities they create would not be a good solution. So even when regulations are created through a political process to manage the problem of externalities, it is likely that externalities will not be entirely eliminated. That is especially true because those who are responsible for creating the externalities will participate politically to minimize the harm they experience from any changes required to minimize the externality.

The excess cost from what people refer to as "internalizing externalities" (in other words, from making the producer of the externalities bear a cost from creating them) may be only in the short term. Over time, the cost of preventing externalities will likely decrease. Innovation can create new ways to accomplish the same goals at a lower cost, and that kind of innovation is likely to happen when many industries need to minimize the pollution they create because of new rules. The initial cost the industry or business may have to bear from regulation is real – and is the reason these actors may fight against regulation – but over time the costs may decrease.

Collective Action Problems

Another reason people don't experience most of the effect of the externalities created by their activities is that environmental problems tend to be collective action problems. In other words, one person's contribution to air pollution is shared by the community that experiences it. The individual polluter might feel the effects of some of the pollution, but most of the effects are felt by others.

Another aspect of what it means to be a collective action problem is that, in many cases, no one person is responsible for creating a problem; each of us contributes a little bit. Fisheries depletion happens because of global or regional fishing; one person's consumption of fish, and even one vessel's fishing, contributes only a small amount. Climate change comes from the actions of many people, all over the world; one person's airline flight, car ride, or home heating forms an unimaginably tiny portion of the problem. This is part of what it means for environmental problems to be diffuse – caused and felt by many different actors – the implications of which are discussed further in chapter 4.

There are implications that follow from understanding environmental problems as collective action problems. Most important is that, for many environmental problems, addressing them requires a lot of entities to change what they are doing. Even if you individually are willing to change, others may not be, and, if they don't change, your contribution will make little difference. If, in order to decrease the problem of plastic pollution, you decide not to take a disposable cup or bag, the environment will hardly notice. The same is true with a fisher who reduces her catch of fish; if others don't do the same, the fish stock will not improve. (In fact, the fish she doesn't take will be there for others to take.) This is one of the biggest arguments for policy solutions to environmental problems. If you need a lot of people to change behavior, you may need rules to ensure that they will.

An additional important element of collective action problems is that whatever benefit is created is shared collectively. If a group of people organize to stop air pollution from a local factory, everyone who breathes the air benefits, even if most of them did not participate in working for that outcome. (The same thing would be true of the people who continued to take plastic bags or cups when others refrained in an effort to decrease plastic waste, or a fisher who doesn't decrease her fishing.) Those people who benefit without contributing are called "free-riders."

It can make sense to be a free-rider. At the individual level, most of us have busy, complicated lives, and making change is hard. If you went to the grocery store without bringing your own bag, the only way to get your groceries home may be to take a disposable bag from the store. The same is true for putting in the organizing effort to make political change. If you don't participate in the protest or lobbying effort in your community to decrease air pollution, and it succeeds, you benefit from the cleaner air as much as those who did give their time and energy to political organizing, and you also have the benefits of whatever else you did with your time or resources.

For businesses the logic is even clearer. Since it can be costly to internalize (or prevent) externalities, bearing that cost when you are not sure your competitors will do so is foolish. Until there is a regulation in place that requires everyone to make the change, being a free-rider on making environmentally beneficial change is likely to be good for business.

Ultimately the issue is that free-riders, or the possibility that there will be free-riders, can make cooperation to address collective action problems extremely difficult. Since anyone can benefit from being a free-rider in the face of successful collective action, fewer people participate in making environmental or political change than should – in other words, most people who want the change, and would benefit from it, don't participate in helping to bring it about. And the fear that not enough people will participate can lead all but the most committed activists to give up on their efforts. After all, if you hold a political rally and only a few people show up, your efforts will be in vain. You receive what game theorists call the "sucker's payoff" – you bear all of the costs and get none of the benefits.[2] Being aware of that risk can make people less likely to participate in collective action.

The environment isn't the only issue that faces collective action problems. Any situation in which action is individual and the effects are collective is a candidate for collective action problems. Students organizing to get better dining hall food is an example of a collective action problem because, if they succeed, the benefit accrues to everyone, regardless of whether they contributed to the effort to improve the collective food. Citizens creating a lobbying day to pass a law requiring internet neutrality benefit no more from their successful efforts than do those who put their attention elsewhere. But additional characteristics of environmental problems, described below, make the collective action problems they face likely to be worse.

Common Pool Resources
The characteristic of environmental issues that makes collective action problems especially difficult is that they are common pool resources (CPR; some economists call them common property resources). These types of issues share two qualities. The first is non-excludability. People cannot be kept from accessing the resources – say, the fish in a lake or the air into which pollution is emitted. Because of that lack of

excludability, almost anyone can contribute to the creation of environmental problems. The effects are also widely shared, and so are the benefits of preventing or fixing the problems – if air pollution is cleaned up, everyone who breathes the air benefits. This aspect is the central one to the creation of collective action problems, discussed above, and can make it difficult to collaborate to solve them.

The second quality causes additional problems: CPRs are what is sometimes called "subtractable" (and sometimes called "rival"). That means that one person's use of a resource (or contribution of pollution to a resource) can make that resource less useful for others. The factory that puts pollution into the air makes the air dirtier for others who want to breathe it. The fisher who takes fish from the ocean leaves fewer fish behind to reproduce or to be caught by others. That is part of what causes the environmental problem in the first place, but the most important aspect of subtractability is that it makes addressing the problem especially difficult. If most factories stop putting pollution into the air but one or two of them don't, those factories can still decrease the quality of the air. If most who are fishing agree to fish less, those who do not change their behavior can simply catch more of the fish that remain. In other words, free-riders don't just make it harder to cooperate (because you know that not everyone is bearing their fair share of the effort to solve a problem), they actively undermine the ability of others to address the problem. In some cases, free-riders can make solving the problem impossible.

Subtractability is a characteristic of common pool resources that other types of public goods problems (such as the operation of lighthouses to warn ships of hazards or the creation of public broadcasting systems) don't have, and that can therefore make them harder to address cooperatively. This logic, again, supports the need for political solutions to environmental problems. Mandating or prohibiting action by all relevant contributors to a problem can be important, since free-riders can undermine any collective solution. But it can be difficult to find political solutions when they would be costly for some of those who would be regulated.

Time and Distance
The potential disconnect between when and where an environmental problem is caused and when and where its effects are felt is another element of environmental issues with important implications for environmental politics. Some environmental problems are felt immediately after they are created, in close proximity to the activities that create them. Indoor air pollution from poorly ventilated stoves, one of the major sources of this type of pollution in poor countries, has these characteristics.

But many environmental issues are experienced distant in either time or space from where the activities that create them take place. Invasive species may become a threat only decades or more after a first non-native species arrives in an ecosystem. Chlorofluorocarbons (CFCs) can cause problems for the stratospheric ozone layer a century or more after they were initially emitted. Some substances, such as greenhouse gases or acid rain, may take time to accumulate in sufficient quantities before major effects are felt.

The same kind of disconnect happens with distance. Much of the plastic that ends up in the garbage patches in the middle of the ocean was used on land, often far from the coasts. Acid rain can occur hundreds of miles from the power plant emissions that cause it. Persistent organic pollutants, such as polychlorinated biphenyls (PCBs) and dioxins, have been found in the blood and breast milk of indigenous peoples in the Arctic far from where these substances were used.

These lags in time and distance matter for several reasons. They can add to uncertainty (discussed in chapter 2) since, if a problem emerges far in time or space from its causes, the connection between cause and effect may not be immediately made. It also means that, by the time a problem is noticed, the behavior that causes it may already be widespread and thus harder to change. And on the side of resolving problems – which is often the stage at which the political process becomes involved – the time between when a behavior stops and a problem is resolved may be quite long, requiring people to take costly action long before the benefit is felt, which politicians may be reluctant to demand. The distance between cause and effect can be politically problematic if the causes of an environmental problem – and thus the location where the costs of changed behavior are felt – is in a different jurisdiction than the location where the effects – and thus the benefits of change – are experienced.

Non-linearities/Tipping Points

Another important element of some environmental problems is that their effects may be related in a non-linear way to their causes. We tend to think of problems as having a clear and consistent relationship between cause and effect: the more CFCs we emit, the more the ozone layer is depleted, and when we stop the emissions the ozone layer will recover. But for many environmental problems the relationship is more complex. There may be tipping points after which changes in natural systems mean that recovery to a previous state is no longer possible. A species may suddenly collapse once a sufficient percentage has been harvested. Enough climate change may cause ocean currents to slow or change direction, dramatically altering weather patterns around the world. More dramatic are feedback loops in which effects compound. For example, as the global average temperature increases, ice melts. Because ice reflects sunlight, when it melts the system takes in more sunlight and warms even more, which causes more ice to melt, and so on. Once this kind of feedback loop engages, it continues to magnify. One of the particular dangers of climate change is that it features many such feedback loops. In other words, more climate change leads to even more (and even faster) climate change. These characteristics of some environmental problems make it more difficult to understand the relationship between cause and effect and increase the urgency of intervening before such dangerous tipping points or feedbacks happen.

Effects of Scarcity: Will We Run out of Resources?
Another important aspect of environmental issues is their intersection with economics: in particular, how people and systems respond to scarcity. Because non-renewable materials are finite, people often express concern that we will use up these resources. Given their importance to our industrial economy, no longer having access to these resources could be problematic. The economist Julian Simon argued that we will never run out of non-renewable resources,[3] and it's worth understanding his reasoning.

Simon made a bet with Paul Ehrlich, a biologist concerned that the world's growing population would lead us to run out of resources, especially those that are not renewable. Ehrlich's logic is easy to understand: as more people use more of these resources, either because there are more people or because the same number of people use more, there will be fewer of them left.

The famous Simon–Ehrlich wager took place in 1980.[4] Ehrlich chose $1,000 worth of five metals – copper, chromium, nickel, tin, and tungsten. He bet that ten years later the inflation-adjusted value of these metals would be higher, which is what you would expect if they were becoming scarcer relative to demand. It is a basic tenet of economics, accepted by both Simon and Ehrlich, that when demand increases relative to supply (in other words, when there is the same amount of something but more people want it, or when the same number of people want something but there is less of it) prices will increase. Simon, who was not concerned that we would run out of resources, bet that the collective price would decrease.

Simon won the wager; collectively the prices in 1990 had decreased. It's worth noting that there were periods of time during that decade in which the prices actually had increased; if the bet had been called at that point, Ehrlich would have won. But the broader trend was on Simon's side, and it's useful to explore both the mechanisms for that and their implications for the possibility of using up non-renewable resources.

It is because these resources become more expensive as they become scarcer (or as demand increases relative to supply) that several other important processes are set in motion. We are more likely to conserve, to substitute, and to innovate because of the increasing cost. For purposes of illustration, let's examine what happens with oil, which includes things such as the gasoline used in most motor vehicles.

When oil becomes scarcer, gasoline prices rise. When fuel costs more, we are likely to use less of it, both as individuals and as a society. Some people might start to carpool to get to work, so that more people are commuting with the same amount of gasoline. Others might wait between trips to the supermarket, so that they use less gasoline per trip. Industry users will try to figure out whether they can become more efficient in their use of fuel, because it costs more. Can they heat or cool buildings less? Run machinery less often? Capture waste heat from mechanical processes to help heat buildings? All of these approaches can fit under the idea of conservation.

We also may be more likely to find substitutes for scarcer, or more expensive, resources. Some people will start taking the bus or subway instead of driving, which could count as both conservation (of fuel) and substitution (of mode of transportation). An electric utility that previously generated electricity with fossil fuels might instead substitute nuclear power or electricity generated by wind. Any of these approaches might have been less convenient or more expensive than using fossil fuels, but once the price of fossil fuel rises the substitution makes sense.

Underpinning all of this is innovation. Because people want to conserve gasoline when it becomes more expensive, they will be more likely to buy fuel-efficient cars, so that gives an incentive to automobile engineers to develop cars that use less fuel. Finding other ways to provide energy that doesn't rely on fossil fuels also makes sense as prices rise – someone has to innovate the ways to get energy from wind or sun and to connect it to a power grid, and it is worthwhile spending the money and effort to do that if the cost of fossil fuels is higher.

Another aspect of innovation involves accessing resources that are more remote and difficult to retrieve. As oil prices rose in the 1970s for a variety of reasons, oil companies were willing to invest more resources in exploration. Since they could earn more per unit of oil they could access, they were able to invest more in finding or extracting it. It was during this period, for instance, when drilling for oil in the deep ocean (such as the North Sea) became cost-effective.[5] It was hard, and dangerous, and required the invention of new technologies to be able to drill deeper and extract and transport the oil. But it was worthwhile for oil companies to invest in this technology, and for inventors to work to create it, because the amount for which each barrel of oil would sell had increased. Likewise, it was worth exploring for oil in places where it was less likely to be found, because, if it were discovered, the payoff would make up for some of the unsuccessful efforts.

We could do the same mental exercise with any non-renewable resource, such as trees or water. As they become scarcer, individuals and society will conserve, substitute, and innovate in ways that both decrease use of the resource and access new reserves of it. So why is it, when we could have more renewable resources if we were just able to leave them be for a while, that we so frequently do deplete these resources, sometimes beyond recovery?

Although demand is the most important determinant of resource prices, these processes of conservation, substitution, and innovation help account for the volatility in prices (and the reason that, in some years of the decade-long bet, Ehrlich would have won the wager he made with Simon). After prices rise and these factors change behavior, prices are likely to fall as the new sources have been accessed and conservation and substitution have decreased consumption. The price trend may be generally upwards, but with notable fluctuations. Some of the conservation and substitution will likely stick, however: once you've bought a more fuel-efficient car, you are likely to use less gasoline, and when you have building materials made out of plastic or aluminum you are no longer in the market for wood.

Were we to get to a point when we were almost out of oil or rare earth elements, or even wood, the price of that resource would be so high that it wouldn't be cost effective to extract or use it. By that point society will have shifted to other ways of providing energy or technological components or building materials. It is in that sense that we will not run out.

The story of innovation and conservation sounds seamless, but it isn't. When prices increase, real people suffer – especially those who are poor. New fuel-efficient or alternative energy cars may be invented, but, for people who simply can't afford to buy a new car, increasing fuel prices will make their lives, at least in the short run, more difficult. The same situation transpires when policies are put in place to decrease the use of these resources through making them more expensive (either directly, by taxing them, or indirectly, by requiring decreased use, which increases the price). Not everyone will bear the same level of personal cost or disruption from these changes. Anticipation of such harms may make those types of policies politically unfeasible. People concerned that addressing climate change will make fuel prices higher may vote or work against taking action, even if they are concerned about climate change, because they fear the short-term costs. There *are* ways that these policies can be made less difficult for portions of the population, and considering those types of policies may help avoid some of the political pitfalls of trying to change environmental behavior.

Even though we may not, technically, use up the last of these resources, there are other reasons to pay attention to them. An important aspect of environmental concern is the ecological destruction that results from extracting resources. Once the most easily accessible sources of these materials have been exhausted, removing them can cause serious damage to landscapes and ecosystems. Mountaintop removal to access underground coal destroys ecosystems and pollutes waterways; mining and processing metals can put toxic chemicals and other dangerous substances into the air and water, among other harms.

Many of these non-renewable resources are themselves the cause of serious pollution, even apart from their extraction. Coal, oil, and gas cause air pollution and are major contributors to global climate change. So even if we are not in danger of actually running out of them, their use causes damage to the environment.

What is Politics?

What does it mean to study politics, and how does that differ from looking at policy? Policy is the specific set of approaches society has chosen (often through governmental processes) to address problems or provide services to people. Politics, on the other hand, is the process within society for making those types of policy decisions. This distinction is important and often elided; we can't understand the rules we create to prevent or address (or fail to prevent or address) environmental problems unless we begin by looking at the different preferences that various sectors of society have for economic or environmental benefits and the political structures through which these social decisions are made.

   Those who advocate for policies to protect the environment often overlook the political aspect of policy: any decision involves a set of tradeoffs between different types of benefits or harms to different groups of people. Politics is the social process of arguing for, and deciding how to make, those tradeoffs. In that process, a simple policy idea may undergo major transformation as political jockeying attempts to carve out benefits for different constituencies. What to economists might look like a simple and effective tax on gasoline is likely to become much more complicated – and less environmentally effective – as exceptions are made for different populations or conditions. In addition, once passed, policy needs to be implemented, and it may face different levels of political capacity to impose, monitor, or enforce rules.

   An essential aspect of this social process of politics is that, while good people may have different opinions about what the best option is, there is no inherently, objectively best way to resolve these tradeoffs. Even if we agree that global climate change is happening, is caused primarily by human activity, and will have problematic effects on people and other species, and even if we understand reasonably well what those types of effects will be, we can have different answers to the question of what, if anything, we should do to prevent it. Those different answers may come because of the direct benefits or costs experienced by different people, because of different levels of risk aversion or acceptance, or because of different ideologies or values.

For example, climate change is caused, in large part, by the burning of fossil fuels (coal, oil, and natural gas). People who make a living, directly or indirectly, from this industry would likely prefer that we continue to use these fuels; after all, they may find themselves without employment if their industry ceases to operate. Countries with large reserves of fossil fuels would have to give up their right to use what is, for them, essentially free sources of energy, a costly decision for them that might constrain other important priorities they have for providing benefits to their citizens. Those people most affected by existing climate change – people who live in low-lying island states already affected by sea-level rise, for instance – are likely to be strongly in favor of reducing fossil-fuel use. (And since climate change is also caused by land-use changes, fossil-fuel industry actors or those who rely on fossil fuels may prefer that, if changes are to be made, they be made by preventing deforestation or changing agricultural practices rather than by restricting fossil-fuel use.)

But even the rest of us, who are less immediately affected by the issue, likely have a range of preferences and opinions on the topic. Are you afraid that transportation will be more expensive or less accessible if we pass rules to decrease the use of fossil fuels? In the short run, that's a realistic fear. And, depending on how those rules are implemented, the burden of those decisions might fall on the poorest members of the population who are least able to afford price increases or new technology. When you consider the characteristics of environmental issues, in which the effects of problems are experienced primarily by people not responsible for causing them, and those effects may be felt far away in time and in space, it's easy to understand why people may not be willing to make their lives harder or more expensive to prevent problems they may not experience any time soon.

The role of science, and scientists (discussed in chapter 2), may be more important to environmental issues than in many other issue areas because of the important role of uncertainty in the creation and resolution of environmental problems. But science and politics interact in complex ways. Even if relevant science is produced and successfully communicated, the political process can turn what appears to be a clear approach into a set of political compromises that undermine the original goal. And science will not save us from having to make political decisions – there is no one right solution to environmental problems, only tradeoffs among options with different advantages and disadvantages for varying groups of people.

What most people focus on when they think about politics are the mechanisms by which these social decisions are made. This is where things such as the forms and processes of governments come in. Do democracies protect the environment better than authoritarian governments do? What effect do different types of governing bodies or the political process of elections have on environmental outcomes? Does the way that laws are created influence the character of rules or the way they are implemented or enforced? These questions are discussed in chapter 3.

Other elements of society also matter in these political decisions. Who are the political actors? Some relevant ones are officially part of the political process. Political parties – how many and of what sort? – can influence what policy options are considered or which ones pass. The bureaucracy is a set of offices and organizations within the government that plays a role in implementing policy; it remains reasonably constant even when elected officials change more regularly. The judicial process – and the courts and judges that comprise it – help decide conflicts among laws or various actors in ways that affect the implementation and enforcement of environmental policy. Non-governmental actors, be they groups of concerned citizens or culpable industry actors, are central players in the politics of addressing environmental issues. Journalism (and, more recently, social media) plays an increasingly important role in channeling information used in political processes. These entities and groups are addressed in chapter 4.

Political decisions become even more complicated when they take place on the international level. In national politics, decisions about environmental policy can compel action from people in that country, whether they support or oppose the policy. But there is no international government that can compel action from countries. Even when these countries get together to create international rules, each country decides, through its own political processes, whether it wants to implement those rules in what might be a global collective action problem. The particular nature of international environmental politics, and its implications for how international rules are created, is discussed in chapter 5.

Finally, chapter 6 brings together what we know about environmental issues and political structures to assess which combination of factors makes societies most or least able to deal with which environmental problems. It also addresses unresolved questions about the workings of environmental politics across countries and political levels.

Ultimately, environmental politics is the process by which society's preferences are created, negotiated, fought about, and translated into a set of policies that influence that society's use of natural resources or creation of pollution. How these political processes play out is strongly influenced both by the special characteristics of environmental issues and by the nature of the political structures in which this set of social decisions takes place.

CHAPTER 2

Uncertainty and Science

A hallmark of environmental problems is that they are frequently caused by activities that we didn't initially know would have problematic environmental effects. Often environmental problems are discovered before we understand what their causes are. In those instances, before we can figure out how to address the problem, as with acid rain or the effects of DDT (a pesticide later discovered to have major ecological and human health effects), scientific investigation works to understand the cause of an environmental problem after it appears. In other cases, as with ozone depletion or climate change, scientists hypothesize that problematic environmental effects will happen before these effects are noticed environmentally. These two models of how scientific understandings of environmental problems develop have important implications for the politics of addressing these issues.

Understanding uncertainty and the role that science plays in the creation, and resolution, of environmental problems is an important starting point for examining the politics of the environment. The role of science and uncertainty – and information more generally – in environmental politics is complicated. The traditional view is that scientific activity helps us resolve uncertainty about how physical processes work, after which the political process draws on that information to make reasonable decisions about resource use or avoiding environmental harms. Even in that context, scientists interact uneasily with the political process. They often see their work as non-political and hesitate to participate in political processes. Scientists and politicians also (metaphorically) speak different languages – for instance, scientists think in terms of probabilities and hypotheses and present findings with caution, while politicians want to deal with certainties. These two groups may not understand each other's efforts to communicate across this divide.

But the role of science is also much more complex. Scientists study what they think to study, and that is deeply influenced by social and political processes. At a basic level, availability of funding, which can be determined by political decisions, can influence what type of research can get undertaken. Even more important is *how* that research is done. Most research (and certainly most policy-relevant research) requires comparisons and categorizations for contexts in which there is no inherently correct way to do so. (What is the best way to measure whether and how much the climate is changing?) This categorical relativity also leaves research open to politicization, especially when science is being communicated to an audience – of politicians or interest groups – that is not especially scientifically literate.

How both this lack of understanding and how scientific information or advice play into the political decision-making process is important. Uncertainty may make political decision-making more difficult, but political decision-making may also transform scientifically sound policy ideas into complicated kludges that bear few of the advantages advocated by proponents.

Relevant to this consideration of the role of science and uncertainty are things such as efforts – particularly important in, but not unique to, current times – to sow misinformation by those who would prefer that political solutions to environmental problems not be implemented. How the public responds to scientific disagreement or uncertainty, as well as the role of motivated reasoning (in which people pay attention to information that confirms their interests), is also an element in understanding how environmental problems are addressed within political processes. At the same time, how people view risk – the likelihood and magnitude of harm in a particular context – adds additional complications to the role of science and uncertainty. Risk is key to social decisions about environmental problems.

Climate change illustrates the issues of uncertainty particularly well. It fits the model of an environmental problem theorized before (rather than after) it is observed in the environment. While it can be advantageous to be able to identify a problem before it is fully manifest, that situation can be more difficult politically: mitigating climate change requires costly social policy in the short term, when few people have knowingly suffered from – or even observed – the problem they're asked to bear a cost to prevent. The science to understand climate change operates in a context that depends on political processes not only for funding but also for defining the types of questions that are considered relevant. It presents a clear instance of politicization of science, with people opposed to addressing climate change funding and promoting their own scientific studies to support their preferred political outcomes. And because there is no inherent way to measure the extent of climate change (or even what combination of effects climate change encompasses), it is possible to cherry-pick comparison dates or phenomena to convey the impression that the climate is not changing notably.

Uncertainty and the Environment

There are often a lot of things we don't know about environmental problems, and that uncertainty influences the political processes for preventing or addressing them. From the very beginning, the nature of environmental issues as externalities suggests that they are often caused without anyone knowing that their actions are contributing to environmental problems. That lack of knowledge contributes to uncertainty, because we may initially create problems without even knowing we are doing so.

There are two different models of how our scientific understanding of environmental problems comes about. The first model involves uncertainty about their cause. Historically, many environmental problems were discovered without knowing initially what the cause was, and a search would then ensue to explain what was responsible for their creation. High-profile environmental disasters following that model include the neurological disease suffered by people in Minamata, Japan, in the 1950s, which turned out to be from mercury in industrial wastewater, and the major human health problems at Love Canal (an area of Niagara Falls, New York), which were ultimately determined to ensue from toxic waste disposed by companies in the area.

Acid rain is another good example of this type of problem. Although people had known for a long time that burning coal caused air pollution (and for that reason people built bigger and bigger smokestacks for emissions), acid rain is often experienced quite distant from where coal burning or other sulfur and nitrogen oxide emissions take place. Swedish scientists in the 1950s and 1960s discovered that the acidity in local lakes was increasing and set about to determine the cause. An early hypothesis that it could result from burning coal elsewhere in

Europe was derided as being implausible; few believed that acidifying substances could travel so far from their source.[1] Eventually sufficient evidence did point to the ability of coal-burning emissions to travel such distances; as the broader process of acid rain was better understood, the world was more willing to undertake changes to address the problem.

In some cases, though, the relationship between scientific understanding of an environmental problem and political efforts to address it works differently. More recently, the discovery of environmental problems, especially international ones, has followed a different model, in which their possibility is theorized before they are discovered in the world. The understanding of ozone depletion is a good example. Ozone is a molecule (O_3) that is a pollutant at ground level, but a layer of it in the stratosphere protects the earth from harmful ultraviolet (UV) radiation. Chlorofluorocarbons (CFCs) are human-made chemicals that were used in refrigeration and foam-blowing; they are extremely stable and are non-toxic, so they were especially good replacements for the toxic and sometimes explosive chemicals that had previously been used for refrigeration.

Basic scientific research showed that CFCs could be broken down by sunlight, and scientists started lab research which demonstrated that, when that happens, it can start a chain reaction that can destroy ozone molecules. Scientists started wondering whether the stability of the CFC molecules meant that they could last long enough to rise to the stratosphere, where there was both an abundance of ozone molecules and sunlight that could break them down. The scientists who theorized this, Mario Molina and F. Sherward Rowland,[2] eventually earned a Nobel Prize for their efforts. So long before anyone had noticed any thinning of the ozone layer, or any effects on earth from such a thinning, scientists suggested that this environmental problem might be happening. They then set about to look for evidence that it was, in fact, taking place, and that it was affecting ecosystems on earth. They found that evidence eventually.

The same kind of model applies to the understanding of global climate change, an environmental problem that happens as a result of what is called the greenhouse effect. This was first theorized in 1827 by Jean-Baptiste Joseph Fourier, who postulated that the atmosphere influenced the temperature of the earth's surface and described what is now known as the greenhouse effect, by which certain gases in the atmosphere trap the sun's radiation from escaping and keep the earth warmer than it would otherwise be. This understanding was further refined through the nineteenth and the beginning of the twentieth century, and international scientific cooperation was undertaken to monitor the atmosphere. The Swedish scientist Svante Arrhenius even explored, at the turn of the twentieth century, the idea that a doubling of carbon in the atmosphere, made possible by human activities, would increase the earth's temperature. These observations, however, went largely unexplored until after World War II, when interest in meteorological data increased on account of better technology and air travel.[3] International climate scientists took up the question in earnest, refining the understanding of the underlying mechanisms and finding evidence of increasing greenhouse gases in the atmosphere, increasing global average temperatures, and the beginnings of a wide variety of predicted global effects.

The science of climate change is extremely complicated, as discussed in chapter 1. Global effects from the phenomenon can be felt centuries after the emissions that contribute to it. Feedback loops and other non-linear effects can make it particularly difficult to predict and measure changes. Current research on global climate change includes the creation of multiple general circulation models – extremely large computer simulations of climatic conditions that attempt to model the effects of changes to the climate. Over time scientists have better understood the mechanisms of the climate system, and the effects they have been able to measure from climate change have become simultaneously more nuanced and more dramatic.

The uncertainty that comes in the context of trying to predict future conditions applies outside of climate science. Another issue where prediction can be complicated is the decline in fisheries. Even when there are decreased numbers of fish of a particular species, it may not be obvious to those fishing for them that they are declining. In some cases that might be because the fishers simply catch other species. In other cases it might be because those who work hard at fishing (combined with the natural tendency for increased technology to make it easier to find and catch fish over time) are able to find fish even when there are fewer of them. It's even possible that, due to some quirks of fish behavior, fish group together when there are fewer of them in ways that make it easier to catch them.[4] Whatever the cause of the uncertainty about how many fish there are, we frequently find ourselves in a situation where scientists studying fisheries predict their decline, while those who are fishing cast doubt on the possibility. (It's also worth noting that the fishers may have a short-run incentive to underplay the possibility of decline for fear that they will have to decrease their catches – at a cost to their livelihoods.)

The types of uncertainty produced in these two different models by which we come to understand the causes and effects of environmental problems have different implications for the politics of addressing or preventing them. In the first model – in which we see an environmental problem and go searching for its cause – scientific information is frequently helpful: it provides an answer to a question people were already trying to resolve about a problem they were already having.

In the second model, it can be extremely difficult to persuade people of a problem whose existence is theorized but the effect of which they cannot yet see. The general population will be uncertain about whether the problem exists at all (in fact, it might not yet exist), and there is obviously at least some level of uncertainty about what harms it will cause. Science can play an important role in leading to additional understanding about what the likely environmental effects are, how bad they will be, and who will experience them.

This type of uncertainty fits in with some of the policy-making difficulties discussed in chapter 1 and elsewhere in this book. Obviously, the best time to prevent a problem is before it has become manifest. But that is also the most difficult time to convince people to do something about it. When a problem has yet to be experienced, you essentially have to convince people to change what they are doing in the present in order to prevent a problem from which they are not yet suffering. Because environmental problems are caused by externalities that don't bear a cost, changing what people are doing to cease producing those externalities will almost certainly initially increase their costs. That means that people have to bear a certain cost in the present to avoid an uncertain (potential) cost in the future. That can be a hard sell even without uncertainty, but uncertainty makes the situation more difficult, because it is at least theoretically possible that the current sacrifice may be unnecessary if the environmental problem wouldn't have materialized after all.

In addition, the public is especially susceptible to disbelief in environmental problems about which there is any hint of uncertainty. Michaël Aklin and Johannes Urpelainen conducted a survey-based study to find out at what level of purported scientific dissent or uncertainty the public was likely to support action to address an environmental problem. Survey respondents who were told about the scientific details of a problem but not given any information on the percentage of scientists who believed in the credibility of that information were reasonably likely to support taking action to address the issue. These respondents also generally believed that the problem described was accurate. Other sets of respondents were given additional information about the percentage of scientists who believed that the results presented were credible. As may not be surprising, the group that heard that only 60 percent of scientists believed the information to be credible were both less likely to believe the information and to support taking action. But that was also the case for groups told that 80 percent or even 98 percent of scientists accepted the results.[5] The fact that a situation where 98 percent of scientists agree – an astonishingly high percentage for those who are used to witnessing the scientific process – made people *less* confident in the results than when they were not told anything at all about scientific belief suggests how malleable public opinion is in the face of uncertainty.

Interestingly, although uncertainty often makes addressing environmental issues more politically difficult, it isn't always inherently problematic to the political process. There are some situations when people may be more willing to act if they are uncertain about the costs or benefits of doing so. Game theory has been used to demonstrate that cooperation to address an environmental problem can be harder when information is clear about who the winners and losers of the negotiation will be.[6] One example of this phenomenon in practice is the negotiation of the environmental protocol to the Antarctic Treaty,[7] in which countries agreed to ban mining on the continent indefinitely. At that point countries were uncertain about the existence of valuable minerals under the ice, and in particular didn't know whether these existed in the areas that countries claimed as their own territory. It is likely that, if any country knew that it was sitting on valuable mineral wealth in territory it had claimed (but was not allowed to act on under the treaty), it would have been unwilling to agree to a mining ban or even to the overall agreement.

Risk and the Environment

Risk is related to, but distinct from, uncertainty in the context of environmental politics. Risk is the likelihood of a bad outcome, such as the possibility of getting cancer from a pesticide or asthma from living close to a waste incinerator, or the danger posed by lead in an urban water supply. The concept of risk, and how people and policy makers consider it, is central to understanding environmental action or inaction in the face of uncertainty. We are concerned about risks that perhaps shouldn't worry us and not concerned about the ones that matter most.

Although there are many ways to define risk, one useful way is the likelihood of an outcome multiplied by the magnitude of its effects should it occur. (Another common definition is that risk is the amount of hazard multiplied by the exposure to it, but that definition makes the uncertainty in risk less transparent.) There are inherent uncertainties throughout any effort at assessing risk, because we may not know the likelihood that a particular chemical will cause cancer or how likely it is that people, once exposed, will contract the disease. Risk assessment, the process of estimating these parameters, is often a part of political discussions about how to address potential environmental problems. In some political processes, formal risk assessment is required as a part of regulation.

One difficulty with risk assessment is that there is evidence that people systematically perceive or value risk in ways that can be seen to be irrational. We have cognitive biases that make it difficult to fully understand uncertainty and risk. Our emotional responses to environmental problems hinder our ability to act.

Much of the study of this phenomenon comes from social psychology and behavioral economics. Some of these problematic risk perceptions come from the existence of uncertainty and how people evaluate probability when there is insufficient information. We tend to systematically undervalue some types of risks and overvalue others. People are especially bad at understanding high-magnitude low-probability events (which is where the most dangerous possibility of future environmental effects exist). We tend to underestimate the likelihood of death due to high-frequency causes (such as asthma) and conversely overestimate the likelihood of death due to low-frequency causes (such as tornadoes).[8] People underestimate the likelihood that they will experience the negative consequences of a potential future problem and overestimate their ability to address these consequences if the event happens.

Another misunderstanding of risk is the idea of "anchoring." People tend to stick to their original assessments of probability even when new information would allow them to update their information and make better predictions.[9] This can be a particularly problematic phenomenon for environmental problems, about which new information emerges regularly (and frequently points to problems such as climate change being more severe and more certain than initially understood).

Among the most important of these tendencies to mis-assess risk is what has been termed "loss aversion." People tend to be more concerned about potential losses than about potential gains in ways that can lead them to make irrational decisions.

In other words, people are risk averse (preferring a safer choice) when the situation concerns losses and risk acceptant (more willing to take a less likely outcome) when the situation pertains to gains. This observation is a key element of what is called Prospect Theory, articulated by Daniel Kahneman and Amos Tversky,[10] the former of whom won a Nobel Prize in Economics for this insight. It has been demonstrated experimentally in many ways. For example, in one experiment, some students were randomly given mugs. Students were then asked to indicate how much they would pay to get a mug (if they didn't have one) or sell it for (if they did). Those who did not have mugs indicated they'd be willing to pay somewhere around $2.50, while those who had mugs were generally not willing to part with them for less than $4.50.[11] In other words, if you didn't have a mug you didn't value it particularly highly, but once you (randomly) had one you were reluctant to part with it. Loss aversion is particularly problematic when dealing with environmental problems, because changing behavior in the present frequently involves giving up something (income, a way of behaving) that you already have.

There is also a problem of what is called "motivated reasoning," in which people fail to take in information that doesn't match their pre-existing preferences or worldview.[12] Not only are people more likely to trust information that fits their biases, they also resist taking in new information that contradicts these biases, sometimes so strongly that their previous (incorrect) beliefs grow stronger in the face of contradictory information.[13] Surprisingly, this phenomenon is more pronounced the more education a person has; highly educated people can be the most resistant to updating their information in light of new evidence that contradicts their preferences. At the same time, people are "overconfident" in their assessments of information and associated risk, and the more likely they are to be wrong, the more likely they are to be overconfident in their assessments.[14]

Risk assessment also has a social component. In principle, risk assessment should apply equally to all of the people to whom the risk applies, but the socio-economic situation they are in causes them to think differently about risk. Studies in the United States have shown that white men, who tend to have the most social and political power, are less concerned about environmental risks than are women of all races and non-white men.[15] In the United States, conservative white men are the most likely to deny the existence of (or human responsibility for) climate change.[16] And one way that risk valuation of ordinary people differs from that of risk assessors is that the general public is especially concerned about risks over which they have little control and which are involuntarily experienced.[17] People without social power may indeed be more vulnerable in these situations, and perhaps it is not a mis-valuation to recognize this experience in how they value risk.

In some contexts, people tend to downplay risks that concern them in what may be a psychological self-defense mechanism. Perhaps surprisingly, people are more likely to downplay risks the more likely they are to experience them.[18] This type of coping mechanism leads us to deny problems or to mistrust information that contradicts our worldview.

Risk assessment is quite a flexible tool: you can assess the likelihood and effects of anything. But the way it is actually applied within the political process can be problematic for environmental issues, since it tends to focus on risks to people. This is understandable in a political context: politicians care about harms to people, because people are the ones who re-elect them and whose interests they need to represent. A focus on people in risk assessment doesn't mean that the potential direct harm to people is the only thing that is calculated, but that other types of effects (such as ecosystem disruption) are assessed to the extent that they would ultimately affect people. But it does mean that those ecological or species effects are calculated only when they create harm for human beings.

Risk assessments may also not focus primarily on all the relevant aspects of risks even for the people who are the center of attention. They tend to focus more on the risks of death than on other types of risks people face (such as long-term disability or illness). How risks are assessed is a choice – which risks, to whom or what, are considered – so risk assessment is capable of being politically manipulated,[19] either intentionally or unintentionally.

All of these tendencies to mis-value risk contribute to the difficulties of addressing environmental issues through the political process. Major environmental problems such as climate change run afoul of many of the difficulties of assessing risk. We tend to overly discount the future, when most environmental effects will take place, and concentrate on the losses we would experience in the present from having to change. Motivated reasoning, combined with anchoring on existing information, makes it more difficult to adapt our expectations to new information that the problem is more severe, and more certain, than scientists used to believe. Psychological defense mechanisms that make us ignore scary problems about which we feel powerless,[20] and other aspects of political processes discussed elsewhere in this book, mean that it may be even harder to address environmental issues politically than other types of social problems where risk plays a less central role.

Scientists as Influencers of Policy

There is no denying that scientists and related experts play a role in the political processes that produce (or fail to produce) environmental policy. They are assumed to be experts and often to be politically neutral, although a more recent tendency in some countries has been to accuse scientists of political bias and try to prevent scientific evidence that contradicts a preferred policy outcome from being presented.

One influential approach to looking at uncertainty suggests that scientific understanding and scientists themselves are what drive successful resolution of environmental problems. This theory of "epistemic communities" was made popular in the context of international environmental politics by the scholar Peter Haas, but it has also been applied at the domestic level.

Haas defined epistemic communities as "knowledge-based networks of specialists who share beliefs in cause–effect relations, validity tests, and underlying principled values, and pursue common goals."[21] In practice, most people who identify epistemic communities see them as scientific or technical experts who may help to resolve uncertainty and frame the relevant knowledge for policy makers. The idea is that environmental problems where such communities exist are more likely to be addressed politically, because these communities help create a shared understanding of the problem and the approach to it.

Science is often embedded in political processes, both at the domestic and the international level. Sheila Jasanoff referred to scientists as "the Fifth Branch" to emphasize the role of science advisors in the American policy-making process.[22] International environmental agreements and the institutions they create (discussed further in chapter 5) frequently create scientific assessment panels as part of their processes.

Although many scholars point to the important political roles played by scientists because of their level of expertise in the environmental problems they research, others point out that scientists are often uneasy with, and ill-suited for, the role of actors within political processes. Scientists and policy makers have different backgrounds and different sets of assumptions and ways of seeing the world, and they may not be able to communicate across this divide. This reluctance may make scientists less willing to do anything other than simply perform research and publish it in academic journals to ensure that policy makers are aware of, and correctly understand, their findings. In particular, scientists may be afraid of compromising their status as disinterested and neutral if they engage in the policy process.[23]

From the other side of this divide, policy makers and others argue that the communication problem of science in the political process comes from scientists not producing "policy-relevant" science. There are efforts at both the domestic and the international level to try to urge scientists to produce information that will be more useful to the policy process. These calls generally suggest that the credibility, salience, and legitimacy of the scientific process need to be strengthened.[24]

Another approach is to expand the definition of what constitutes science or who is consulted as scientists – the expansion of what is sometimes called "citizen scientists" or "civic science." The idea is that regular people often have expertise or knowledge, sometimes beyond that to which the formal scientific process may have access. One particularly important element of this expansion is how science integrates the knowledge of indigenous people, especially when it comes to biodiversity loss or climate change. People with a close connection to ecosystems over time may be best placed to document changes.

The Relationship between Science and Politics

The role of scientific advice within the policy process – in other words, the *politics* of science in environmental decision-making – is even more complex than the role of science in understanding environmental issues. The conventional thinking about the relationship is that science provides the basic information about an environmental problem and resolves any uncertainty that exists, and then, once we fully understand it, the political process takes over and creates policy to address it.

There are many ways in which that concept doesn't reflect reality. Science and politics are not as separate as many think. Science is often represented as apolitical, as scientists study questions that interest them with the methods and approaches that are most useful for doing so; information from this "ivory tower" process then makes its way into the political process. But this description mischaracterizes the scientific process in several ways.

First, what "science" even studies in the first place is influenced by political processes. Scientists think to examine things that are in the public eye, and thus issues of public concern help direct scientific analysis. Likewise, scientific research requires funding, and so research projects that are able to garner funding, because they are of greater concern to those who control access to finance, is the research that gets undertaken. In many places, much funding for science comes from public resources, so the priorities of governments can influence what research happens. To take an extreme case, in the United States for much of recent history, public funding of research into the effects of guns or marijuana has been prohibited, making it extremely difficult to study those topics.

More important, though, is *how* those issues are studied. What is the case for all research to some extent is particularly true for studying the environment: there is no set baseline against which it makes sense to compare whether things have improved or become worse. Studying most things requires categorization and comparison, and there are no inherent categories. Is the climate warming? Compared to when? Our current average temperature is noticeably warmer than it was 125 years ago, but it is somewhat cooler than it was 125,000 years ago and dramatically cooler than it was 55 million years ago, when dinosaurs roamed the earth.[25]

What is even relevant to look at when we're examining whether the climate is changing? Temperature? Measured how? Should we look at the average temperature or the extent of temperature extremes? Measured where? Is temperature more useful, or a more reliable predictor, of what we want to know when we measure on land, in the ocean, or in the atmosphere?

Should we include rainfall? Storminess? Drought? One of the main arguments for conceptualizing what some call "global warming" as climate change is that many different aspects of the climate are likely to be affected by an average warming. But, if some areas become wetter and some dryer (and some warmer and some cooler), a measurement of the average may miss those results, which nevertheless matter dramatically for the places that experience those changes.

The same thing is true for comparing effects. If you wanted to look at the different environmental effects of the way we generate electricity, for instance, you would discover that some (burning of coal and oil) contribute to climate change (and burning coal also causes acid rain), some (hydropower and, to a lesser extent, wind) have effects on fish and wildlife, some (nuclear) cause dangerous forms of waste, and so on. These environmental effects are different, and there's no inherent way to combine them into one measure or to prioritize one over another. How much water use equals how much contribution to climate change? How much toxic waste is worth a reduction in particular air pollution? There's no clear way to prioritize those tradeoffs. Some people who conduct "life cycle assessments" attempt to put all environmental effects onto one scale, but doing so requires making a judgment about how much of one effect should be traded off against another.

The bigger rejoinder is that there is no right answer to those questions. The philosopher Dale Jamieson refers to this phenomenon as "categorical relativity"[26] to point out that any effort to categorize requires choosing a starting, or comparison, point when no inherent point exists. You can make an argument for studying any of these comparisons or starting points. Honest scientists do their best to focus on ones that they think will be most useful or, even, sometimes most socially relevant.

For instance, there is recent evidence that people are more likely to be affected by an increase in the average low temperature than in a general increase in average temperature; these higher "low" temperatures can also have broader ecosystem effects by increasing the number of frost-free days in a region and thus affecting disease vectors, pest survival, or crop health.[27] Understanding the extent of that change would be useful. And it may also be the case that some species are more sensitive to variation in temperature than to warming per se,[28] something that would be missed by simply looking at a change in average temperatures.

But this is another place where politics enters the equation. With all these different types of effects, or different ways of measuring, it's easy to pick and choose the information to present based on the argument you want to make. Those actors who have a preference for what the political outcome will be – particularly those who engage in activities that might be regulated were efforts made to address the environmental problem – have the ability to pick and choose the information they present to make a case against taking action.

Those opposed to action on climate change are particularly prone to picking and choosing their comparison dates to try to make the case that climate change is not severe, has paused, or is not happening at all. They often do so by picking an unusually warm year (like 1998) and then a recent unusually cool year, and drawing a graph that compares the two. Because there is large variation in annual temperature (some of it driven by atmospheric cycles such as the La Niña/El Niño phenomenon that are scientifically well understood and not directly related to climate change) it can be reasonably easy to find a recent set of years that, when compared, show the temperature to be remaining stable or even falling, despite an overall trend of increasing warming.[29]

The generally low level of scientific literacy among the population (and among policy makers as well) makes this problem even more severe; people do not know how to interpret scientific information and do not think to question information presented. It is why they are likely to misunderstand how a major snowstorm or cold snap does not negate the presence of "global warming" and thus makes them susceptible to the misleading information presented by those who prefer inaction on climate change.

There can be social influence in how science is conducted more generally, even without intention to deceive. Categories, comparisons, or even questions depend on what we think to ask, and that depends on the social constraints that influence how we see the world. Decades of medical research was done on men, in part because it didn't occur to those undertaking the research that women might, as a category, be differently affected by medical issues or procedures.[30] The same is true for people of different ethnic backgrounds.

This research can even be influenced by what we expect to find. Scientists studying ozone depletion had to program their instruments for a range of expected outcomes. Early computer programs had been set to regard drops outside of a set range as errors and discard the readings, because many random errors caused by equipment failures or atmospheric interference could quickly swamp the useable data. Since no one expected ozone measurements to be nearly zero, extremely low measurements were assumed to be outside the realm of useful data, and actual ozone depletion detected by scientific instruments was initially ignored – or never even seen – by the scientists working with the data collected.[31]

Another way that scientific research can be affected by people concerns what Jamieson calls "agency uncertainty."[32] For environmental issues especially, what people decide to do affects the nature of the environmental problem, because it is people's actions that are contributing to those problems. But how people – or governments – behave may itself be affected by the scientific estimates about the problem. Jamieson gives the analogy of traffic reporters predicting backups on the road to the beach on a sunny day. Listeners who hear that the prediction may decide not to go to the beach, and the traffic jam may thus be avoided. Does that mean that predictions were incorrect? Or that their estimate changed behavior? In other words, we don't actually know how people will behave, and their behavior may be affected – for good or for bad – by information about environmental problems. That makes understanding the likely parameters of these problems more difficult.

A worse situation comes from ignoring the role of agency in fixing environmental problems. The Danish statistician Bjørn Lomborg is among scholars who frame themselves as "skeptics" on environmental issues. Lomborg somewhat disingenuously argues that even if environmental problems are real we don't have to worry about them, because when others have emerged in the past they have ultimately been solved, and we will therefore find fixes to our current ones.[33] But it is precisely because of the role of human agency that we were able to do this: people became worried about environmental damage, changed their behavior, and pressured their governments to act, and that is what led to the improvement.

It is for this reason that the Intergovernmental Panel on Climate Change (IPCC) reports its predictions in what it initially called "scenarios" and more recently has framed as "Representative Concentration Pathways." These are different predicted trends of ways human populations (including their governments) may act over time with implications for climate change effects. Many different scenarios have been used, but they include such elements as speed and type of economic growth, type of main energy source, extent of population growth, globalization vs. localization of the economy, and other social and economic factors that could affect global greenhouse gas emissions.[34]

Scientific Advice Meets Politics

Even the most careful and relevant scientific advice is rarely translated directly into policy outcomes. That is neither surprising nor necessarily problematic; sometimes it represents distrust of science, but it may also represent a political choice about priorities. Political approaches to addressing fisheries depletion and climate change provide useful illustrations.

In the case of fisheries management at the international level, international organizations regulating fishing generally have scientific commissions that make recommendations about the level of fishing a stock can sustain in a given year. Actual catch limit decisions are made by a political process of voting by countries after scientific advice has been given. Across these organizations and over time there is a lot of variation in the relationship between scientific recommendations and the level of catches agreed on through the political processes, but catch limits are generally set higher than scientific advice recommends.[35] Some of that represents the difficulty of negotiation at the international level, where countries do not have to comply with international rules (as discussed in chapter 5), and rules may be watered down to lure in reluctant states. But some of it also involves decisions by states about their domestic fishers who would be hurt by decreased catches and the political opposition from fishers, processors and others concerned about the economic effects of decreased fishing.

Domestic political processes may take simple scientific (or social scientific) recommendations and make them more complicated in order to address concerns other than environmental ones. Policies to address climate change that involve carbon (or even just gasoline) taxes are likely to affect poor people more than rich ones, and so exemptions or refunds may be carved out, or subsidies added, to avoid creating these extra harms on more vulnerable populations. And, of course, other actors who will be affected by such a policy are likely to argue against it as well, and so exceptions may also be granted for politically powerful actors. The policy that gets created may therefore be much less simple than proposed and achieve less environmental benefit, at a higher cost, than the initial recommendation would have done. Which is to say that politics will generally transform scientific advice when implementing it.

The Special Case of Climate Change Denial

A more problematic situation is the outright rejection of scientific information. Climate change is real, it is happening, and much of it is caused by human actions over recent history. That is the consensus of the vast majority of climate scientists in the world. Over time, these propositions have become clearer and the evidence has become starker. There are many things about which uncertainty still remains in climate science. The exact details of the effects, as well as their locations and magnitudes, remain uncertain, and climate models, though they are improving, cannot yet understand everything about the workings of the climate system. But the underlying science is, at minimum, clear, and the understanding that climate change is happening, is problematic, and is caused primarily by human activity is widely shared in the scientific community.[36]

And yet, in some places – particularly, although not exclusively, the United States – the popular understanding of climate change has at points gone in the other direction. Fewer people now are persuaded that climate change is a real phenomenon, or is caused by people, than at earlier periods of time. This "climate denial" is not accidental and plays an important role in the politics of climate change. Understanding it is essential to understanding the role of science in environmental policy more generally.

Climate change faces more misinformation and attempts to mislead people than do most environmental issues. Business interests which feared that, if people took climate change seriously, the resulting regulations would harm their business model and bottom line have contributed financing to help sow misinformation about the basic details of the environmental problem. In the United States, oil companies donated to think tanks to encourage the production of reports – of dubious quality – to cast doubts on the scientific consensus on climate change or the role of people in creating it.[37]

With scientific evidence not on their side, those opposed to addressing climate change have also attempted to prevent information about it from being used in regulatory processes. In the United States these efforts have happened primarily in state (sub-national) governments. For example, in 2015 and subsequent years, the governor of Florida prohibited the use of the terms "climate change" or "global warming" in official communications.[38] North Carolina in 2012 banned the use of sea-level rise projections in coastal policy.[39] Under the Trump administration, the Department of Agriculture created a list of words related to climate change to be avoided in official communications.[40] Because of categorical relativity and the low level of scientific literacy, these efforts to sow confusion about the extent of scientific understanding of climate change has had real effects on political willingness to address it.

Science and Social Decisions

One of the most pervasive assumptions about the relationship between science and policy is that simply resolving uncertainty – getting the science "right" – can make political decisions easy. A tweet from the astrophysicist Neil deGrasse Tyson, in which he argued that his preferred political process is one in which decisions "shall be based on the weight of evidence,"[41] represents this type of thinking.

Those who focus on simply allowing science to resolve our political differences over environmental problems often point to the need to improve the "science–policy interface" or "communicating science." In other words, if only we could get scientists both to produce more policy-relevant science and to communicate it better to those engaged in making policy, science would be more likely to produce policy outcomes that are better for the environment. Scholars who take this approach focus on the concern that scientists can be reluctant to engage in political processes, or that they speak different languages than policy makers (making their results sound less conclusive than they may actually be), and that they may study abstract or non-policy-relevant aspects of an environmental problem and thus fail to produce the type of information the political process needs. This framing is not just misguided but actually dangerous: it assumes that there is a technically right answer to what we should do about an environmental problem, and that all we need to do to reach it is to get the scientific information right.

That perspective misunderstands politics. A political process, imperfect and problematic though it may sometimes be, is an effort to navigate tradeoffs among different social values. Politics is about deciding what we should do and about how those decisions should be made. Often (though not always, as discussed above) those decisions are made easier with a full range of information. For example, we may be concerned about the possible negative human health effects of pesticides or fertilizers. At the same time, though, those fertilizers and pesticides help increase crop growth in the short term and thus make it possible to feed more people. Farmers make the informed decisions that their labor costs will be lower, and yields and profits will be higher, if they use them. There may be as many downsides to reducing the use of these chemicals as there are upsides.

Navigating these tradeoffs becomes even more difficult when different sets of populations experience the costs and the benefits of environmental problems or what is done to address them. Farmers who use fertilizers may bear the full benefits of the increased crop growth, while the resulting algae blooms happen hundreds or thousands of miles downstream, affecting fishers or tourists. Chapter 5 discusses the additional difficulties that result if these populations are in different political jurisdictions, but even if everyone that participates on all sides of a problem is in the same country, you may have a political disagreement among farmers (and those who represent them in government), who would suffer economic harm if they were no longer allowed to use the fertilizer in question, and fishers or tourists (and those who represent them in government), who are harmed as long as the practice continues.

Assuming we know the full costs and benefits of using, or ceasing to use, these fertilizers, we could theoretically do a dispassionate calculation as to which option leaves us collectively better off in terms of the economic costs or benefits. But does that necessarily mean that that would be the right option to choose? Does it matter how many people are harmed by each option, or how involuntarily? Is it even realistic to put a price on harms (such as illness or death) that aren't clearly economic? Making these decisions requires making choices about what (or who) is more or less important to prioritize, and having more information is not a substitute for making those choices.

Conclusion

Scientific uncertainty can be both a cause of and an impediment to solving environmental problems. Basic scientific research can help us understand what is causing environmental problems and can predict problems that have not yet appeared. The latter kind are becoming more common and can be the most difficult to address, both because they tend to have long time-horizons and because it can be difficult to persuade people (or political systems) to take action to address a problem that has not yet become apparent.

In working to reduce uncertainty, scientists can play an important role in the politics of the environment. It must be understood, though, that scientists do not work in a vacuum apart from politics. What they study, and the way they study it, is more influenced by the social and political context than even they may realize. And their work, in large part because of the inherent social construction of science, is subject to politicization. Efforts to improve what some call the "science-policy" or "research–policy interface"[42] may result in better communication or more policy-relevant science, but they cannot change the fact that science alone cannot resolve policy decisions or make political choices.

We need to understand the role of science in environmental politics not only to appreciate that science will not save us from having to make social decisions; it is also to be aware of the way in which science and uncertainty are deployed by those engaged in struggles about what those social decisions will be. People attempt to use science as a way to avoid fighting about the underlying social priorities, when it is simply an input into politics rather than a way around it. Framing information, picking and choosing points of comparison, magnifying uncertainty to play up the extent to which a problem might not actually happen, or might not be worth preventing, are all strategies used by organizations, businesses, or groups of people opposed to taking action to address environmental problems.

But this framing of science is also used by those in favor of environmental protection, attempting to avoid having to make the difficult social or economic choices that legitimately exist in addressing environmental problems. Ultimately, acknowledging that environmental problems are real (and frequently understood well enough to consider addressing) is important, but so is accepting that there are tradeoffs in deciding whether or how to address them. As with most other types of politics, there will be winners and losers, in both the short and the long term, from preventing or fixing environmental problems. Science can be helpful in clarifying who those actors will be, but it will not rescue us from having to make social choices that will affect different groups of people differently.

CHAPTER 3

Political Structures

Examining the politics of the environment requires an understanding of the types of governing structures within which political decisions are made. Rather than examining the political processes of any one specific country, this chapter looks at the different types of political structures within states and the implications they have for political decision-making about environmental issues, with examples drawn from different countries that have those political structures. If politics is about navigating competing social preferences, then the context in which societies argue or compromise about these decisions, and the rules those contexts impose, should influence the political outcomes.

A first-cut distinction is the difference between democracies and authoritarian (or other forms of non-democratic) regimes. Although it might seem obvious that democratic states would have better environmental records, both the theoretical framework and the empirical evidence are somewhat more mixed. This chapter examines the theory and empirics suggesting the environmental advantages and difficulties of each broad political type. Democracies may be more likely to reflect the interest their populations have for a cleaner environment, but their political processes can be contentious and slow. Authoritarian governments may be less likely to be concerned about environmental issues overall, but when they decide to act they have the ability to make swift change: witness the ability of China just before the 2008 Beijing Olympics to suspend the operation of factories and dramatically curtail traffic, with significant environmental benefits (reductions of sulfur dioxide emissions by 60 percent and of carbon monoxide and nitrogen dioxide by nearly 50 percent),[1] actions that would not have been feasible in a democratic regime.

Imperfections in democracy, such as corruption or inequality, as well as transitions to democracy, influence the environmental politics of the countries in which they apply. Within democracies, different types of political structures affect the way the environment is addressed politically. One difference is that between federal systems, in which multiple subsidiary units (e.g. states or provinces) have the authority to make political decisions within certain realms that cannot necessarily be overruled by the national governing process, and unitary systems, in which the national government is the primary decision-making authority and can overrule lower-level political decisions. Among other important variations, federal systems allow for different political decisions about rules across different regions of a country, which has both advantages and disadvantages for addressing environmental issues. In the United States, for instance, California has historically had much more stringent air-pollution regulations than the rest of the country. That kind of difference can serve as both an experiment – what are the economic implications of stricter requirements? – and an incentive for businesses such as automobile manufacturers, who might choose to produce to the higher standard so that their products can be sold in any state. At the same time, the unevenness in regulatory standards nationally may leave some parts of a country with much worse environmental conditions than others.

Also important are the differences between parliamentary systems and presidential (or separation-of-powers) systems, as well as those between disparate types of parliamentary systems. Since environmental interests (as discussed in chapter 1) are generally diffuse, thinking about how different electoral systems create more or less opportunity for representation of these types of interests within elected governing bodies is key to understanding how the environment will fare within political processes. Presidential systems create a greater number of points of access, but that can make policies harder to pass and easier to overturn or block once they have been passed. Parliamentary systems may make it easier for smaller parties to gain some representation and even to have a chance of being part of a governing coalition. That helps account for the role of green parties in some European systems. Different ways in which parliamentary elections are run (single-member districts versus proportional representation, for instance) create different types of opportunities.

The Role of Democracy

It seems obvious that democratic countries would have better environmental records than non-democracies. After all, if people in general prefer better environmental conditions, and democracies are governments designed to represent the will of the people, political structures designed to represent the population should reflect those interests.

It turns out that the record is much more complicated than that, and also that it is difficult even to figure out the role democracy plays in environmental outcomes. There are several reasons for that. First, there are many other confounding factors. A variety of things we know lead to better environmental conditions and policies – such as greater wealth, higher levels of international interaction, and greater educational attainment, along with market economies – also correlate with the level of democracy in a country. So if you see countries with all of those characteristics, it's hard to know which factor is most responsible for contributing to the environmental outcomes. Second, while there are advantages that democracy brings to the addressing of environmental problems, there are also disadvantages; and the same can be said of authoritarian governments. Thus while some characteristics of democracies might lead them to be more willing or able to address environmental problems, other aspects complicate their efforts.

Let's begin by defining what it means to be a democracy. Most countries in the world are now democracies, though they may be imperfectly democratic, and there are many different ways to define democracy. Common approaches focus on the procedures or institutions of democracy.[2] Elements of these that are frequently discussed include the idea of free and fair elections, in which citizens have the right and ability to vote for their preferred candidates or policies. These processes usually imply the idea of both open competition among candidates and multiple political parties. Also sometimes made explicit is the idea of majority rule both for elections (the candidate with a majority of the votes wins, although in some cases rules allow for candidates or parties with a plurality to win) and for the governing processes themselves.

Another framing is the idea of liberal democracy; in this case, "liberal" signifies not any kind of ideology but, rather, a set of protections: individual rights, most particularly the rights of the minority, are safeguarded. If being outvoted in an electoral process could enable some portions of society to lose essential rights (or, at the worst, their life or liberty), they would not be able to continue to participate fully in the democratic process. Other types of freedoms are also guaranteed in this kind of democracy. Things such as a free press and freedom of speech, the right to assemble, and other related rights are seen not only as beneficial in their own right but as central to the guarantee of a fair and open electoral process. If citizens are not able to say what they think about candidates or preferred outcomes, and if news outlets are restricted in what they can report, it is not clear that elections can be held fairly. When we consider the relationship between the environment and democracy, it is most frequently liberal democracy to which people are referring.

There are a variety of ways scholars measure democracy. For the most part, "democracy" isn't a binary condition. It's notable that even countries that most scholars would regard as non-democratic have some of the trappings of democratic processes, such as elections. There may simply be so many restrictions on who is eligible to run for office and who is allowed to vote that these places aren't considered to be democratic. Freedom House, an independent organization based in the United States focusing on freedom and democracy, puts out an annual index that measures how "free" countries are (designating them as free, partly free, or not free). The index includes each country's respect for political rights (based on ten indicators) and civil liberties (based on fifteen indicators) and results in a total score out of a possible 100 points. Freedom House also categorizes which states are considered to be electoral democracies, prioritizing countries that have met minimum standards concerning competitive multiparty systems, universal suffrage, regular elections conducted by secret and secure ballots, and open campaigning with fair access by political parties to the media.[3]

Another commonly used rubric is the Democracy Index, compiled annually by the UK's Economist Intelligence Unit (EIU). The specific details of the index change annually but include measures across five categories: "electoral process and pluralism, civil liberties, functioning of government, political participation, and political culture,"[4] collectively producing a rating out of ten points in each category, averaged for an overall score. Both of these indices change their measurements regularly, so a country's trajectory is not immediately comparable year to year, but trends can be seen, and in any given year a country's level of democracy can be compared with others in the index.

These two approaches measure slightly different things and so reach somewhat different – though generally compatible – conclusions about the countries they examine, as evidenced in table 3.1, which compares Freedom House and Economist Intelligence Unit ratings for an assorted set of countries.

Other measurements exist. The Polity Index, for instance, focuses primarily on institutional criteria within central governments, measuring six aspects that include "executive recruitment, constraints on executive authority and political competition."[5] There have been four generations of this index, each measuring democracy in a slightly different way. Ultimately, although each of these large-scale efforts at categorizing democracy emphasizes or measures a slightly different set of aspects, the ratings they produce are quite comparable, and some scholarly studies use multiple measurements when analyzing effects of democracy.

Even within countries that are considered to be democratic there can be restrictions on who in the population is able to vote or run for office. Historically, women were frequently not given rights to vote, and the same has been true of people from ethnic minorities. These days, we would not consider places with these restrictions to be democracies, but the history of most of the world's democracies includes these types of restrictions. Full democracy, therefore, is an aspiration and not simply a binary measurement.

Table 3.1 Comparative democracy measurements				
	Freedom House score (out of 100)	Freedom House rating	Economist Intelligence Unit score (out of 10)	Economist Intelligence Unit democracy rating
Brazil	75	Free	6.97	Flawed democracy
China	11	Not free	3.32	Authoritarian
Denmark	97	Free	9.22	Full democracy
Ghana	83	Free	6.63	Flawed democracy
Japan	96	Free	7.99	Flawed democracy
Kenya	48	Partly free	5.11	Hybrid regime
Nigeria	50	Partly free	4.44	Hybrid regime
Portugal	96	Free	7.84	Flawed democracy
Russia	20	Not free	2.94	Authoritarian
Saudi Arabia	7	Not free	1.93	Authoritarian
Turkey	31	Not free	4.37	Mixed regime
United Kingdom	93	Free	8.53	Full democracy
United States	86	Free	7.96	Flawed democracy
Uruguay	98	Free	8.38	Full democracy
Zimbabwe	31	Partly free	3.16	Authoritarian

Source: Freedom House, Democracy in Retreat: Freedom in the World 2019 (Washington, DC: Freedom House, 2019), https://freedomhouse.org/report/freedom-world/freedom-world-2019; Economist Intelligence Unit, Democracy Index 2018: Me Too? Political Participation, Protest and Democracy (London: Economist Intelligence Unit, 2019), www.eiu.com/public/topical_report.aspx?campaignid=Democracy2018.

A Detour into Authoritarianism

One way to consider the role of democracy is to begin by looking at countries that aren't democratic – what are frequently called authoritarian countries. While there is no one definition, authoritarian governments are generally those with strong central power, a lack of meaningful political participation by citizens, and restrictions on individual freedoms.[6] One of the primary reasons scholars started to examine the role of democracy in environmental conditions is that some authoritarian countries had particularly terrible records on the environment, and that these conditions seemed traceable to characteristics of their governments.

Primary among these was the Soviet Union (USSR), an authoritarian one-party state – essentially an agglomeration of Eastern European territories and Russia – that existed between 1922 and 1991. It had a terrible environmental record, featuring pollution of air and water, issues with waste management, and serious nuclear contamination from its weapons development program during the Cold War with the West.[7] Its central planning and prioritization of industrial development and economic growth, along with political repression that prevented citizens from having a say in the governing process, resulted in dreadful environmental conditions.[8]

Why are authoritarian governments so likely to create bad environmental conditions? They often prioritize quick economic growth and are concerned more with economic benefits despite the potential environmental downsides that accompany them. Most authoritarian governments are in poorer states, undergoing industrialization; for that reason they are particularly concerned about economic growth.[9] That is especially likely to result in bad environmental conditions because these types of states are insulated from the public pressure that might persuade them to address environmental issues even when doing so would decrease the short-term profitability of industry.

Authoritarian governments need to focus on remaining in power and thus are concerned with providing quick benefits to their populations. The short time-horizons of electoral politics – the desire of officials to be re-elected – is one reason that democracy does not protect environmental interests as well as we might expect. But the pressures in authoritarian governments are even more stark. While the state apparatus in authoritarian states may have more control over the societies they govern, widespread unhappiness is more likely to threaten the large-scale structure of society – in other words, a transition away from the current governing structure would be a complete and dramatic loss of power for the authoritarians governing, unlike a political party within a democracy which might simply be out of power for a short while but still have electoral chances in the future.

Environmental problems within authoritarian states may even help cement the authority of those states. For instance, it can be argued that China's efforts to deal with its dramatic air pollution problems (as discussed below) using distinctly authoritarian tools further advances its rule. In other cases, environmental concerns are seen by authoritarian rulers as "safe" outlets for citizen action. For example, in Iran, state-sponsored (or state-monitored) civil society NGOs have been permitted to provide basic environmental services to the society in part as a way of responding to serious environmental problems so that these problems do not ultimately undermine governing authority. At the same time, channeling participation into state-sanctioned environmental organizations also decreases political pressure on other aspects of the governing authority of the regime.[10]

In other cases, though, those very environmental problems

(and the political outlet they provide) may also be responsible for democratizing pressure. Environmental movements within the former Soviet Union were one of the few forms of domestic protest not immediately shut down, and for that reason these organizations formed a reasonably safe way for those who opposed the regime on many different grounds to organize, as scholars such as Jane Dawson argue.[11] This environmental action is seen as having played an important role in the ultimate downfall of the Soviet Union.

There are some advantages that authoritarian governments have for addressing environmental issues, however, and some scholars argue that ecologically minded authoritarian states have certain abilities to address environmental problems that democracies don't. The idea of ecologically minded authoritarian states is sometimes referred to as "authoritarian environmentalism." Bruce Gilley defines this approach as "a public policy model that concentrates authority in a few executive agencies manned by capable and uncorrupt elites seeking to improve environmental outcomes,"[12] arguing for its potential superiority as a policy-making model.

Some scholars, who others call "eco-authoritarians," are making a theoretical argument rather than an empirical one. They do not necessarily find ecological advantages from existing authoritarian regimes but, instead, argue that some aspects of environmental problems might benefit from being addressed through ecologically enlightened governments that are willing and able to act without democratic interference from the populace. This argument itself previews the downsides of democracy for addressing environmental problems, discussed below. These eco-authoritarians point to the necessity of restricting some of the human freedom to pollute or to overuse resources, given the incentive structures (discussed in chapter 1) that lead even individuals with perfectly reasonable individual motives to cause collective ecological damage.[13]

When people do point to empirical evidence of the advantages of authoritarianism, China is the primary example. China has made greater investments in renewable energy than most countries and, despite having enormous coal reserves, has been making important strides in transitioning away from coal, something democracies such as the United States have recently faced much more difficulty in doing. It is precisely the authoritarianism of China that can allow large-scale social engineering that makes possible these types of energy transitions, or other big infrastructure projects, without the possibility of collective social protest that would accompany such policies in most democracies.[14] Singapore is another not-full democracy (Freedom House classifies it as "partly free")[15] to whose advantages in environmental protection some scholars point, particularly its use of green spaces and its infrastructural development as a green city.[16]

Both political entities combine these green elements with serious environmental downsides, however. Despite Singapore's parks and infrastructure, it has major problems with air and water pollution, as well as deforestation and species extinction. China's successes at temporarily remediating air pollution may have been made possible by its authoritarian rule, but that pollution can arguably be attributed to development unmediated by public concerns. Judith Shapiro points out that China's voracious development not only creates environmental problems within its own borders but also contributes to deforestation, resource extraction, and the generation of pollution worldwide.[17]

Mechanisms for Democracy's Environmental Influence

The starting point for the influence of democracy on environmental conditions or commitments is the idea that people prefer better environmental conditions and that democracy represents the will of the people. Therefore, the logic goes, democratic countries should focus on creating better environmental outcomes. It is useful to explore the processes by which this influence could happen. Different scholars concentrate on different aspects of democracy that might lead to better environmental outcomes, though these aspects are often interconnected.

One approach examines the role that individual freedoms play. The ability of individuals to access information, and to organize, allows them to influence governmental action. This holds true as well for the ability of environmental organizations to gather and present information and lobby the political process for the outcomes they prefer; it is the ability of these types of groups to engage with the political process within democracies that allows them to press effectively for environmental policy.[18]

A second approach focuses more on the institutional mechanisms of democracy as the explanation for its predicted relationship with better environmental outcomes. For example, there is evidence that improvement in the robustness of democratic institutions decreases the level of deforestation in developing countries.[19] Some of this general institutional emphasis is on the role of civil society and the judiciary (both discussed in chapter 4). Of central importance is the occurrence of meaningful elections and a legislature that represents the interests of the population.[20]

Democracy is also understood to play an important role in the environmental Kuznets curve (EKC). The EKC describes a known relationship between a country's per capita wealth and

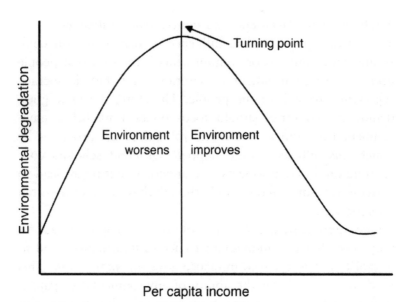

Figure 3.1   Environmental Kuznets curve

Source: Anh Hoang To, Dao Ha, Minh Ha Nguyen, and Duc Vo, "The Impact of Foreign Direct Investment on Environment Degradation: Evidence from Emerging Markets in Asia," *International Journal of Environmental Research and Public Health* 16/9 (2019): 1636–50.

its environmental behavior, such as emissions of some type of pollutants or resource use. For many environmental issues, societies at low levels of wealth are not major emitters of pollutants. As the wealth of a society increases, it is more likely to engage in environmentally polluting (or resource-degrading) behavior. But at a certain level of wealth the curve turns around, and with increased wealth comes decreased environmentally damaging behavior.

There are a variety of factors that account for the turnaround point of an EKC. Economists generally point to changes in the economy that move away from heavy industry and towards service. Off-shoring plays a role, too, as some of the more polluting economic activity that benefits a country takes place elsewhere. Also important are domestic rules that constrain polluting activity. This is where the most direct effect of democracy is seen: those rules come about because of domestic political pressure that is most prevalent in democracies.[21] It is not surprising that EKCs are most frequently seen on issues that have direct and obvious local effects on a fairly short time-scale; these are the sorts of issues that are likely to cause local mobilization. This kind of political pressure can be seen as playing a role in some of the more economic aspects of an EKC as well – offshoring and changes in industry are likely, due in part to political pressure for better environmental conditions that is more effective within democracies. Some studies have found that democracy improves long-term EKC relationships for the types of pollutants (e.g. sulfur dioxide)[22] and resource use (e.g. deforestation)[23] that EKCs typically involve.

The Relationship between Democracy and the Environment
As expected, when we look at actual environmental conditions there are some aspects of the environment that fare better in more democratic countries. Foremost among these are procedural issues. Democratic governments are far more likely to make environmental commitments, including joining both international environmental agreements and international environmental organizations and implementing their rules.[24] They also are more likely to create domestic environmental regulations. On a few measures of actual environmental conditions – most notably emissions of methane and chloro-fluorocarbons (CFCs) and land protection – countries fare better the more democratic they are.[25]

On the other hand, there are a number of factors on which democracy leads to worse environmental conditions. Some types of air pollutants, such as volatile organic compounds

(VOCs), are worse under democracy, along with certain types of soil erosion. More prevalent are environmental conditions that show a mixed relationship with democracy across different studies: deforestation, greenhouse gas emission, and some water pollution.[26]

It might not be surprising that democracy doesn't inherently correlate with better environmental outcomes. Democracy is slow; deliberations among representatives, taking into account competing perspectives on what is best for society, do not happen quickly. Most democratic political systems are actually designed to work slowly – the deliberation and debate is part of what it means to be a democracy. But environmental problems may need quick action, and democracies aren't good at reacting quickly. Moreover, the kinds of political power exhibited disproportionately by business and industry in democracy (and discussed further in chapter 4) can persuade politicians not to create the regulations necessary to prevent environmental harm by these actors. The characteristics (discussed in chapter 1) that make environmental problems difficult to address collectively – including the role of externalities and disjunction in time and space between the cause and effects of environmental problems – can be magnified in democratic processes, where politicians are rewarded for benefits experienced immediately by the populations they represent.

Democratic Transitions

Although there is some evidence of the long-run advantage of democracy for some environmental issues, particularly the existence of environmental policies, the *transition* from authoritarian to democratic regimes can be tumultuous, including for the environment. For example, deforestation tends to be greater in transitioning countries (or newly democratic countries) than it is in either democracies or authoritarian countries.[27] In Kenya, for example, the increased competitiveness of elections created incentives for political actors to allow use of forest resources as a way to curry favor with the electorate.[28]

There are several reasons why countries undergoing transitions to democracy may experience worse environmental conditions during that transition. First, democratizing countries lack robust institutions, so, even if they have some of the trappings of democracy, they do not have the underlying structures to ensure their beneficial effect. One of the major environmental benefits of democracy is in the implementation and enforcement of rules, and new democracies particularly lack this ability.

Second, the legacy of a lack of democratic participation may stifle citizen action, even when it is nominally allowed under the democratizing regime. Civil society, seen as central to the benefits that democracy brings to environmental issues, tends to be weak at the beginning of democratic transitions. This is not coincidental; the legacy of authoritarian repression of citizen action leaves such groups weak and reluctant at the beginning of a newly democratic regime.[29] Or rights to information or to organize might exist, but people might not be aware of them. In other cases (such as in Japan and in the former Soviet Union), non-governmental environmental organizations seemed energized at the very beginnings of democratization but "failed to mature," and many organizations disappeared from the political process. Miranda Schreurs attributes this weakness to the restrictive regulations on citizen organizations left over from authoritarian times.[30]

Finally, the transition to democracy typically features widespread political and economic upheaval accompanied by uncertainty, and uncertainty can be especially problematic for beneficial environmental behavior. It can cause movement backwards along an EKC, especially if the transition causes wealth to decrease.

Ironically, though, one of the effects of the economic difficulties of transitions to democracy is that it can improve environmental conditions on the ground. This can be seen, for example, with the transitions in Eastern Europe and the former Soviet Union. One of the implications of the shutting down of centrally planned (and environmentally disastrous) heavy industrialization was an immediate improvement in air quality and a dramatic decrease in emissions of greenhouse gases. The economic decline that followed this transition was also responsible for a continued diminishing of emissions of the types of pollutants associated with economic activity and energy use.[31] The combination of economic decline and newly democratic institutions probably accounts for more environmental improvement than would otherwise be the case.

Imperfections in Democracy

Because democracy is more of a spectrum than a binary condition, states that are nominally or even functionally democracies may still contain some elements that are not fully democratic. When considering the history of democracy in the last several centuries, it is clear that the goal of representative government with full and equal participation of its citizens (not to mention the question of who is eligible to be a citizen) is in most places a work in progress. To the extent that these non-democratic aspects themselves lead to worse environmental policies or outcomes, they can serve as further evidence of the usefulness of democracy for the environment. In other words, working to improve how democracy works can be useful in its own right but also useful for its ability to lead to environmental protection. All of these factors can also exist in authoritarian governments and contribute to making their governing processes less responsive to environmental concerns than they would already be.

Corruption

One of these imperfections is corruption. The World Bank defines corruption as "the abuse of public power for private gain."[32] It can comprise such things as governmental officials taking bribes to refrain from enforcing existing policies, requiring bribes or favors to give needed permissions for various activities, or other exchanges of benefits that persuade officials to look the other way when otherwise illegal activities happen. More systemic corruption within political systems can include embezzlement or misuse of official funding sources or manipulating political outcomes or processes to gain individual wealth or influence, among many other possibilities. Corruption is not unique to democracies – authoritarian governments often rely on corruption to function – but when it exists within democracies it can potentially undermine any environmental advantages democracy may bring.

Corruption is hard to measure because it is intentionally hidden: corrupt officials are generally breaking the law. Scholarship examining corruption frequently uses the Corruption Perceptions Index created by the non-governmental organization Transparency International. Rather than attempting to characterize corruption directly, the organization surveys citizens of most of the countries in the world as to their perceptions and experiences of corruption within their countries. The answers are then aggregated to a score out of 100, with the highest-scoring countries the least corrupt. In the most recent survey, even the top-scoring countries do not make it into the highest ("very clean") category, suggesting that there is some level of corruption within all countries. The 2018 index lists

Denmark and New Zealand as the least corrupt countries and
Somalia, South Sudan, and Syria as the most corrupt; it also
calls out the United States and Brazil as "countries to watch,"
reporting increased corruption in both.[33]

Not surprisingly, most scholars studying the issue have found
that corruption leads to worse environmental outcomes.[34]
When those who have the most resources (most frequently,
polluting or extractive industries) have the greatest access to
the policy process, environmental conditions will be worse
than they would be otherwise. Increasing the reach of these
actors in corrupt contexts, by augmenting the ability they have
to influence policy or their ability to avoid being held to existing policy, will only make environmental situations worse.[35]

The effects of corruption on environmental outcomes are
likely to be uneven. Some environmental resources are more
economically valuable than others and thus more likely to be
the subject of corrupt deals for overexploitation. Similarly,
economically powerful industries that pollute are more likely
to be able to avoid environmental rules in corrupt countries
than companies that bring in less wealth.

There are also indirect ways that corruption can influence
environmental outcomes. Corruption decreases the wealth of
a society, since money changing hands goes into the private
incomes of corrupt officials rather than into socially beneficial economic activity. So, to the extent that environmental
outcomes improve (eventually) with the wealth of a society,
corruption will interfere with that improvement, making it
happen at a higher level of wealth than would otherwise be
required.[36]

On the other hand, there are some circumstances when corruption can improve environmental outcomes. Because it gives
additional influence to those who already have some kind of
influence, it can increase the influence of environmental lob-

bying groups. One group of scholars found that ratification of the Kyoto Protocol to the Framework Convention on Climate Change was more likely the more environmental lobbying activity that there was, but also that this effect is stronger the more corrupt a country is.[37]

Corruption magnifies other effects. The more open a country is to trade, the more stringent its environmental policy is likely to be, but that effect is even more pronounced the more corrupt a country is.[38] If corruption has diminished the stringency of environmental policy, a corrupt country has room for more improvement, so the benefit from trade openness will be magnified.

Corrupt governments overall are thus likely to be worse, in most cases, for the representation of environmental interests, as well as for democracy more generally. Corruption can be hard to overcome, but doing so is likely to be beneficial on multiple political and environmental dimensions.

Inequality

Inequality, based on wealth or on other social factors such as race, is a major problem in many political systems and can undermine the extent to which democracies fully represent or protect their citizens or other residents. Inequality generally, and particularly in terms of access to full participation in political processes, is not only a problem in its own right but also likely to decrease the responsiveness of political processes to environmental problems.

There are a number of elements of inequality and ways that they may play a role in environmental politics. Economic inequality has been particularly well studied. One way that economic inequality within a society is measured is by the Gini coefficient. This measurement of the income distribution within a state is on a scale from 0 to 1 (or sometimes represented as

0 to 100 percent). A society with a Gini coefficient of 0 would be perfectly equal (everyone in the society would have the same income). For example, Denmark's Gini coefficient is .26, Canada's is .31, and that of the United States is .39; more unequal countries are Brazil at .51 and South Africa at .63.[39] This framing can be used to examine how societies that are more unequal in terms of wealth or income compare environmentally to those that are more economically equal. In general, societies with greater inequality (but similar in other demographic ways) face worse air pollution, energy consumption, deforestation, biodiversity loss, and access to safe water and sanitation than societies with more equal income distributions.[40]

It should not be surprising that inequality within political systems generally leads to worse environmental outcomes, as the vast majority of empirical studies demonstrate. Environmental degradation is more likely to be caused by those who are powerful and wealthy. The less access those without power have to the political process, the less able they are to use that process to protect themselves, and society in general, from environmental degradation. Inequality also worsens collective action problems (discussed in chapter 1) by raising the opportunity costs of already difficult collaboration for the poorest members of society. In other words, organizing politically is hard, and if you have fewer resources it is even harder. Societal-level inequality can also reduce a state's willingness to bear the (near-term) cost of environmental protection: poorer segments of society may be less willing to prioritize addressing environmental problems (and bearing the costs to do so) over other social or economic issues.[41]

Inequality also can affect environmental Kuznets curves, most likely shifting them to the right – in other words, under conditions of inequality, it takes a higher overall income level to move to a turnaround in which further increases in wealth lead to decreases in environmental degradation. It is local environmental issues (rather than global ones) that are most likely to be affected by inequality, since those are the ones to which the EKC is most likely to apply. In particular, the voices of those most affected by local issues are the ones most likely to be silenced by inequality.[42] The same logic (and empirical evidence) applies in the case of environmental problems with immediate health consequences, which are more likely than those with longer-term health implications to be affected by inequality.

There may be some environmental benefits of inequality. Some approaches to environmental conservation, such as land protection, impose a cost on society and may be difficult politically. There is now increasing attention to the fact that land protection often comes at the expense of the most disadvantaged people.[43] Inequality may thus make it easier to protect land at the expense of underprivileged members of society. This effect of inequality is more prominent the less democratic a society is otherwise. One study examining 137 states showed that greater economic inequality led to less protected land in highly democratic societies, but greater levels of economic inequality led to more protected land in less democratic societies.[44] Other studies have found that more unequal societies may do better at protecting land, perhaps because they are more willing to impose costs on their most disadvantaged populations.

Overall, there may be countervailing effects of economic inequality that can make it difficult to tease out its effects. Since poorer populations have lower ecological footprints overall,[45] unequal societies may decrease some aspects of their environmental effect by having more poor residents than their aggregate income level would otherwise suggest. Nevertheless, the environmental downsides of inequality are likely to coincide with other negative social and democratic effects, suggesting important advantages to working to make societies more economically equal.

Environmental Injustice

There are other forms of inequality that matter environmentally. What is frequently called "environmental injustice" refers to the fact that, within political systems, people who are the most socially vulnerable in a variety of ways are also likely to suffer the most from environmental degradation and have the least political voice in its resolution.

The Environmental Justice movement began, as such, in the United States, with the observation that, whatever the overall level of environmental quality, African Americans, as well as people from other minority groups, suffered more from environmental degradation within communities. There are a variety of proximate causes for this disproportionate effect: members of minority groups are more likely to reside in urban areas with heavy industry or in proximity to highways or other transit corridors,[46] and also near hazardous waste facilities. The literature on environmental justice is quite clear: across many societies (and when considered internationally as well), people who are socio-economically disadvantaged also experience more environmental harms.[47]

It can be difficult to disentangle the effects of income, race, and other social characteristics because they often intersect: in many populations, racial minority communities are also the most impoverished. And environmental justice scholarship finds important effects of poverty on the extent to which a community bears a disproportionate environmental burden.

But an important insight from environmental justice scholarship is that race has an independent effect on the environmental conditions experienced by a community and that it can itself account for the broader economic status through a history of systematic discrimination.[48] The same is true of other social factors, such as language spoken or citizenship or immigration status. People who are discriminated against for social factors other than wealth then also have – because of that discrimination – less access to wealth and social position. If in addition they are exposed to worse environmental conditions, the injustice is magnified.

Even when this relationship exists, it can be difficult to disentangle the process by which it transpires: often it is not as simple as outright discrimination.[49] After all, industry would not be likely to choose to operate in minority communities simply because of racism; there must be some economic advantage to siting locations. One possible mechanism by which this disproportionate effect happens is that there is an economic rationality to placing industrial activity or other hazards in these communities; the cost of land, for instance, is lower, or there is access to transportation and roadways. This economic rationality can belie an underlying social inequality: the reason that land is cheaper may well be because the people who reside in the area have less social and economic power; they live in this place because they do not have alternatives. That lack of economic power and alternatives is almost certainly itself traceable to racism.

Similarly, it may be the case that an area transitions demographically because of an environmental hazard. When an environmentally damaging activity is located in a particular area, that area may initially look demographically like any other. But once an area suffers from environmental damage, those with the social or economic means to leave may go elsewhere, and property values decline. The people who move in (or who are unable to leave) are the ones with little social status or economic opportunity.

Even more likely is that environmental hazards end up located in areas where the population has little political power. Populations that are not able, because of any number of social or economic disadvantages, to mobilize to prevent environmentally damaging activities in their areas are presumably more likely to experience these harms. The anticipation of the lack of mobilization can also play a role: polluting industries may not even consider setting up in privileged neighborhoods because they are reasonably certain of pushback and of a political system responsive to the people who live there. But when people in an area may lack citizenship, skills in the dominant language, social status, or economic resources, it may be a safe bet that they will lack the resources to prevent such harms before they happen.

The reason it is important to understand the underlying causes of environmental injustice is to figure out how the political system will, or won't, be able to ameliorate or prevent it. The likelihood that lack of access to political power can result in worse environmental outcomes for communities is both an important observation when studying environmental politics and one more reason to work to improve the situations of communities that suffer disproportionate environmental harm. One important lesson from the environmental justice movement, though, in the United States and elsewhere, is that populations are nevertheless able to mobilize when they suffer unfair harms. The environmental justice movement both in the United States and worldwide is becoming a powerful force for political and environmental change.

Types of Electoral Systems

Within democratic systems, there are different ways that the mechanisms of elections and governing are set up that have important implications for how environmental interests are likely to be politically represented.

Federalism

Most countries in the world have what is known as a unitary system of national government, which means that the primary laws for a given country are created at the national level and apply uniformly across the entire country. But a subset of countries have federal systems, in which regional governments (often called states or provinces) have some degree of auton-omy in passing legislation that pertains only to their political units, and there can thus be policy that differs across sub-units of a political system. In practice, the extent of federalism can vary – in other words, how many powers and which ones are held by the regional sub-units – but in practice we tend to label systems as being either federal or unitary. It's also worth noting that most countries have smaller administrative sub-units that might even be called provinces or states; that doesn't necessarily mean that all of these are federal systems. There are approximately twenty-five federal systems in the world, the best known of which are the United States, Canada, Germany, Australia, India, and Brazil.

It would not be surprising if there were systematic differ-ences between federal systems and unitary systems in terms of the environmental policy approaches they take, but there are competing logics for what the effects should be. As a starting point, it is fairly clear that we should expect more variation overall in levels of environmental protection within federal systems, because environmental rules, or economic rules that have environmental implications, are likely to be one of the governance areas reserved for the sub-national units. If states or provinces are allowed to make their own environmental rules, it should be obvious that these rules would have more overall variation than they would in unitary systems where the national government makes rules that apply to the entire country. As a starting point, there are likely to be variations in levels of political power among actors across different sub-national jurisdictions that might lead to different types of political outcomes relevant to the environment. It's also likely that conditions differ in the various regions of a country, and that local politicians are more tuned to the interests – be they environmental or economic – of their local constituencies, leading to more variation.[50] The question, though, is whether on average the level of environmental stringency will be higher or lower in federal systems than in unitary ones.

There is a logic for the decreased stringency of rules in federal systems. States or provinces want to compete for business and industry, and the best way to do that is to keep environmental standards low. This is a version of what people call the "race to the bottom" hypothesis. Because environmental problems are caused by externalities, regulating to prevent the creation of those externalities comes with a cost for businesses, which will need to act in different ways. So it will be cheaper to operate in a jurisdiction without environmental rules than in one with rules. In a multi-jurisdictional area, it would make sense for a business to set up operations in the area without rules. But because these units are competing with each other for economic activity – if a business is established in one location, that location gets all the economic benefits of employment and other income from it – areas may be reluctant to take on a higher level of regulation. Some scholars make the argument that this variety in regulatory levels is a good thing, allowing for local choice in how to value the tradeoff between economic and environmental wellbeing.[51]

On the other hand, it's also possible that federal systems could, on the whole, lead to more stringent environmental regulations. In part this increased stringency could come from the ability of individual areas to experiment with higher levels of regulation than might be politically possible on a national level; if – as is frequently the case – they turn out not to be especially costly and to have strong environmental benefits, other areas might then choose to adopt them. This is what David Vogel famously called the "California Effect," because in the United States California has frequently adopted higher environmental standards.[52] Other states, which might not have been willing to take these on from the start, come to believe that the rules might benefit them as well, and take them on, leading to collectively stronger environmental rules than would have been the case without the ability of California to make its own rules within a federal system. Within the United States, an especially strong federal system, there is little evidence of federalism and its associated variation leading to a decrease in environmental regulation.[53]

Ultimately, the primary effect of federal systems is that environmental rules are likely to vary more than they do in non-federal systems. This variety, though, is likely to lead to some rules being more stringent and some weaker than would be the case in a unitary system.

Parliamentary versus Presidential Systems
Another important difference across democratic political systems that could have implication for environmental politics is the difference between parliamentary and presidential (or separation-of-powers) systems. This difference concerns the relationship between the executive (the policy leadership) and legislative (the lawmaking) branches of government. In a presidential system, the head of the executive branch (the president) is elected entirely separately from the legislative branch. That can – and often does – lead to a situation in which the titular

leader of the government comes from a different party than the legislature. Enacting rules within this kind of system involves interaction between the different branches of government (hence the concept of separation of powers) in which each has the ability to "check" the other. In the United States, for instance, laws passed by Congress (the legislative branch) must be signed by the president to become effective; the president can veto them, which keeps them from being enacted. But the legislature can override a veto by a supermajority vote (two-thirds of both the Senate and the House of Representatives). So each group has the ability to block action by the other.

In parliamentary systems, the prime minister is simply the principal among the members of the legislative branch. Elections happen in a variety of ways (whether voting is for individuals or for parties), and the head of the party that wins a majority of seats in the elected legislature becomes the prime minister. In some political systems, such as that of the United Kingdom, this outcome is the most common. If the party with the greatest number of seats does not secure an outright majority, it still gains the right to try to form a government by creating a coalition with other, smaller, parties to get to the point where it collectively has a majority within the governing coalition. (In some cases, if no other party is able to form a majority coalition, it is possible for the party with the largest number of votes to govern without one.) In many parliamentary systems coalition government is the norm. Legislation that can pass the parliament becomes law without the possibility that it will be blocked by the executive branch.

Other elements follow from these distinctions: in presidential systems election timelines are usually fixed in advance, with elections happening at specific intervals. In parliamentary systems there may be time periods within which an election must be called, but elections can take place at any period before then and for different reasons: the government can (strategically) decide when to hold an election, or if the governing coalition falls apart an election may take place because of a lack of confidence – which can result in an actual vote of no confidence – from some members of the governing party or coalition.

There are some ways that parliamentary systems are likely to be better able to represent environmental concerns within a political decision-making process.[54] The first is that, if the executive is interested in – or commits to – addressing environmental concerns, the unitary aspect of government allows the prime minister to enact that environmental agenda. There tends to be fairly high levels of party discipline – members of a party will vote similarly on a given issue. The government can also be held to its promises, because the party that controls the government should be able to get legislation it favors passed. So if the government made promises about the environment to get elected, the voters can count on the party to deliver or may punish the party electorally if it doesn't (and if the voters care enough about that issue).

Types of Parliamentary Systems

Other characteristics within parliamentary systems may matter. How many parties predominate in a system can be important. A number of factors contribute to how many parties are serious contenders within an electoral system, with the election rules as well as culture and historical path dependence playing a role. In general, parliamentary systems are more likely to have a greater number of parties than presidential systems, but that can still vary across states. The main reason that parliamentary systems are likely to have more parties than presidential systems is the likelihood of coalition governments. Among parliamentary systems, coalition government predominates. That means that even small parties have the chance to play an important role in a governing coalition by helping a plurality party get to the majority it needs to form a government. The greater the number of viable parties, the more likely that green, or environmental, parties will be a part of the political scene.[55] (The role of parties, and green parties in particular, is discussed further in chapter 4.)

Beyond that observation, though, there is not necessarily a clear relationship between the number of viable parties in a parliamentary system and environmental outcomes. Having more parties in general tends to lead to less stability (and a greater likelihood that government will change hands more frequently), which can be problematic for the representation of environmental interests.[56]

How those parties are elected also varies and can have implications for the representation of environmental (and anti-environmental) interests. Because the structures across parliamentary systems can differ in so many ways and the implications for environmental politics are not always clear, only some of these distinctions are mentioned here, and only briefly. One major distinction is whether the system works by proportional representation or majoritary/plurality. Proportional representation can be accomplished through a variety of types of actual voting rules, but conceptually it is designed to give parties the same proportion of seats as they have proportion of votes. Sometimes it may involve voting for a party rather than for an individual. Under this approach, parties maintain a list of candidates, and how many votes the party gets determines how many, and which, of those candidates are elected into the legislature. Majoritarian/plurality systems generally involve voting directly for candidates, though there are also systems that mix elements of the two.

Proportional representation has some advantages for the representation of environmental interests (even among parliamentary processes) because it tends to result in an increased diversity of perspectives among representatives; these types of systems allow a greater number of smaller issues (which is, practically, how the environment is generally characterized) to matter more for the formation of coalitions.[57] In addition, in proportional representation systems candidates do not need to appeal to the entire electorate (or the entire electorate within a district) in order to be elected. This approach thus allows for parties to appeal to smaller interest groups by including candidates responsive to their issues, and it may be possible for these candidates to be elected even if not all voters support their positions.[58]

The flip side of those observations is also true: most scholars find that majoritarian systems have downsides for environmental policy. First, it can be harder for smaller parties (e.g. green/environmental parties) to make it into government. Second, if electoral rules essentially penalize small parties, there is little incentive for majority parties to adopt the positions of these parties, because they are less likely to be an electoral threat.[59] Duverger's Law is an observation in political science that elections using plurality-voting measures (such as those in which the first candidate to get the largest vote is declared the winner within single-member districts) are more likely to lead to two-party systems, whereas proportional representation is more likely to lead to a multiparty system.

For the sake of completeness, it is worth mentioning that there are hybrid systems – France has one, as do Armenia and Namibia, among others – in which there is both a president, who serves some functions of a separate executive branch, and a prime minister alongside a parliament. Not surprisingly, these systems can be expected to have some of the environmental advantages and disadvantages of each system, especially depending on the local variety of electoral elements exhibited.

State Capacity

A more nebulous aspect of political structure is the ability of a political system to implement and enforce its decisions. That capacity can influence which policies are realistic to advocate, or even whether it makes sense to try to create political solutions to environmental problems. Theda Skocpol defines capacity as "the ability of states to implement official goals, especially over the opposition of powerful social groups, or in the face of difficult economic circumstances."[60] Many aspects contribute to – or detract from – this capacity, some of which, such as corruption or inequality, are discussed elsewhere in this chapter. Capacity may look different in a democracy than in an authoritarian state,[61] though it can matter in both; authoritarian states that can govern by diktat or fear may need less social buy-in to implement their plans.

Wealth, both aggregate and per capita, is another likely contributor to state capacity for effective environmental action. Having governmental resources makes a big difference for the environment, and not only in terms of the relationship between wealth and environmental performance discussed elsewhere in this chapter. Implementing environmental policy in some cases depends on expensive infrastructure. The ability to provide access to sanitation and clean water, for instance, requires an astonishing level of physical piping laid throughout a geographic area, together with processing plants that can distribute or process water or wastes. Providing that infrastructure is costly and legislatively complicated. (The tendency of governments without that capacity to agree to privatize those services only underscores the role that capacity plays.) Infrastructure matters in other ways. If there are no roads in areas where illegal activity may be taking place, it can be difficult to know that it is happening or to apprehend those engaging in it. (On the other hand, that kind of infrastructure can also lead to increased opportunity for destructive environmental activity; roads in the Amazonian forest make logging and taking of other species more accessible.)

Also important in capacity are other elements that can be more difficult to quantify. The bureaucracy (discussed in chapter 4) includes people who monitor environmental conditions, calculate the benefits or costs of action, and enforce policies. Having a cadre of educated, technically competent people dedicated to this role matters – as does the governmental resources to compensate them adequately so that this career choice makes sense for people with the relevant background. Ultimately, then, the ability of a government to implement its environmental decisions depends on more than simply its political structure.

Conclusion

Understanding political structures is central to understanding the politics of the environment. Despite the efficiency of authoritarian political structures, democracies give the most leverage points for environmental politics (as well as providing other important social protections). How elections happen – what type of representation, chosen within what electoral rules – influences who ends up in office in ways that are likely to matter for environmental politics.

Democracies, despite their bias towards the present and the overrepresentation of politically powerful actors (who are also the most likely to benefit from environmental degradation), have important advantages for environmental action. Better environmental conditions are likely to benefit the general population of an area, and the types of transparency and representativeness that democracy is supposed to represent should make environmental protection possible.

Within that context, the most important lesson for environmental politics is to pay attention to the role played by political structure and electoral rules. One way to think about these elements is in terms of how many "veto players" an electoral system introduces.[62] These are the places in a political system where potential action can be stopped by actors who oppose it. Presidential systems, for instance, have a veto point – the requirement that two different branches of government agree in order to implement a policy or ratify a treaty – that parliamentary systems generally lack. Similarly, parliamentary systems (such as those with more than two significant parties) that generally result in coalition governments also introduce veto options compared to those with one or two dominant parties. Also potentially relevant is the number of legislative bodies – it is obviously harder to pass rules that have to go through two legislative bodies than through one. There are good reasons to believe that the more veto points an electoral system has, the harder a time it will have making environmental policy.[63]

On the other hand, these veto points also provide opportunities that can be exploited. Federal systems can be seen as providing more veto points because the sub-national units have the ability to engage in their own environmental politics, and that may result in a lack of uniform national policy. While that may be true, it does allow more environmentally minded states or provinces the ability to move forward on environmental action even when the country as a whole is not willing to do so, and that sub-national action may make national action more likely over the medium term. Multi-member districts or proportional representation systems may lead to a greater number of parties and therefore more veto players, but at the same time they might allow more marginal views to be represented.

Knowing the political structures of a country is thus central to understanding the context in which its environmental decisions take place and important for determining the best entry points for influencing its environmental politics. While the structures may not be amenable to quick change, understanding and working within their constraints for environmental progress can yield important results, and ignoring them will lead only to frustration and ineffective efforts of political action.

CHAPTER 4

Political Actors

If politics is the process by which societies make choices about what they value and the tradeoffs they are willing to accept to pursue those goods, those choices and values are defined and argued over by actors within the political process. Factors discussed in other chapters – political structures and the characteristics of environmental problems, among others – influence the strategies these political actors follow and their ability to gain their preferred outcomes.

Politicians themselves want to be re-elected; that focuses political representatives on outcomes that benefit the most influential of their constituencies within reasonably short time-frames. Political parties to which these actors belong can also influence how political decisions are made. With the occasional exception of so-called green parties, most political parties have a range of issues they are concerned with, and even parties that label themselves "green" tend also to address issues apart from the environment. Which parties hold, or are vying for, political power can have important implications for what decisions are made about the environment.

Whatever political decisions are made need to be implemented and enforced, and the bureaucrats who oversee these efforts play a role in that process that is often overlooked. Because they tend to persist in power for a longer period of time than the elected officials who create the policies, they may provide a source of consistency amid political change. Courts are also key players in the politics of the environment. How independent a judiciary is, and whether its most central members are elected or appointed, varies across political systems. While courts may not be directly involved in the creation of environmental policy, they play a key role in the process of implementing or upholding policies that are created and, as such, are relevant to environmental politics.

The actors of civil society – the area apart from the formal process of governing, where groups of citizens try to influence the policy process – are central in advocating for or against environmental action. Citizens' groups and other organizations (as well as scientific and academic organizations) are often the entities pushing for environmental protection. Industry and business are often responsible for the activity that causes environmental degradation, and political solutions to these problems frequently impose a cost, at least initially, on businesses. For that reason, business actors may resist environmental rules and may argue that increased costs to their operations that would result from action to prevent environmental harm would impose broader costs on society. Or they may participate in political processes to try to influence the form of regulation, seeking to minimize harm to their operations or gain advantage over their competitors. Business activity on the environment can be influenced by the political process in other ways too; businesses may self-regulate or take on other measures to protect the environment in order to persuade political actors that there is no need to regulate their behavior.

How the public becomes aware of environmental issues is shaped by how these issues are presented in the media, and quirks of how journalism works can augment uncertainty or create misunderstanding. Because of the important role of the media, it becomes a site of contestation in itself, with different groups trying to influence how environmental issues are presented based on what their underlying objectives are. The current age of social media, and even the rise of "fake" news (as well as the tendency for some people to disbelieve information they don't like), complicates the relationship between the media and environmental politics.

The interaction of these political actors and their interests matter too. For instance, political opportunity may arise when there is an overlap between the interests of business and environmentalists. These observations help us understand political outcomes, especially in conjunction with an understanding of the political structures discussed elsewhere in this book.

Politicians

In democracies, politicians face regular elections. Depending on the political system, elections may happen on a predictable schedule or be called within a permissible time-range. In general, politicians want to be re-elected. They have what is referred to as the "incumbency advantage," factors that make it easier to be re-elected once in office than for new politicians to replace them. That includes the actual resources that being in office conveys: the ability to send out official mailings, or to create a record of accomplishments as a reason for re-election, or even factors as simple as name recognition that comes from familiarity. If you lose an election, your opportunity to make policy disappears, so re-election tends to be the first priority of politicians.

This process has important implications for how politicians see environmental issues and the steps they take to address problems. On the one hand, environmentalism has been called a "consensus movement," supported by majorities of most populations and opposed by few.[1] That preference applies not only in developed countries but in the developing world as well.[2]

People indicate that they want to protect the environment, and even that they are willing to bear some cost to do so. And, unlike with such contentious issues as abortion or capital punishment, where there are organizations making strong arguments for both sides of the issue, there are no interest groups arguing in favor of polluting the air or wiping out endangered species. But although people generally support environmental action, that support tends to be fairly shallow; it is almost no one's first priority. Ion Vasi characterizes the public's attitudes of concern about climate change, for example, as "a mile wide and an inch deep."[3] And other priorities people have – such as economic wellbeing – may be compromised in the short term by action taken to address or prevent environmental problems.

The need of politicians to get re-elected has implications for the time-horizons they face. They have to deliver benefits to the electorate within the two, four, six, or however many years before the next election. The trouble with environmental problems is that solutions take time and, in the short term, often cost money. Even if cleaning up polluted air is a major benefit, closing down or regulating a factory that is causing air pollution has short-term costs for some of the politician's constituents. And the benefits may not be felt for years, long after the election has taken place. In that context, efforts to address environmental problems by government may come down to which group is able to put the most pressure on politicians, and, with that pressure, the implication that future electoral or governing success is in doubt if the politicians do not deliver what the public demands. Even in states that are not full democracies, civil society groups may be able to bring enough pressure to bear to influence policy outcomes.

Political Parties

Political parties put forward candidates for election. Parties tend to be somewhat stable over time in terms of the broadscale goals they represent. Different political systems have different types and numbers of parties (see chapter 3 for the relationship between political structures and numbers of parties), and thus parties have varying ranges of political positions. One important difference between parties and interest groups (discussed below) is that parties generally have positions on many different issues. A left-wing party may, for instance, favor environmental protection, along with a strong welfare state, regulations on international trade, national healthcare policies, and opposition to foreign wars. The environment is likely to be only one of the issues on which it takes a position. Citizens deciding to join, or vote for, particular parties may be drawn in by only a subset of the party's interests. For that reason, parties can be an imperfect avenue for environmental action.

One major exception is the existence of "green" parties in some electoral systems. Although they often take positions on a variety of matters, their signature issue is the environment. Some of the first green parties were formed in Europe in the 1970s and reached their strongest position in those countries in the 1980s. But there are green, or environmental, parties in countries around the world, and environmentally oriented parties were forming in Australia and New Zealand in the early 1970s as well. Green parties are gaining new power in the developing world.

Not surprisingly, green parties have had greater electoral luck in parliamentary systems, especially those with electoral rules that support a greater number of parties, than in presidential systems. At this point only a small number of prime ministers have come from green parties, and even then only as part of coalition governments rather than from an outright electoral victory of their party: Indulis Emsis of Latvia was the first green party member to become prime minister anywhere; he held that position for nine months. Moana Carcasses Kalosil, of the Green Confederation, was briefly prime minister of Vanuatu (from 2013 to 2014), as well has having previously been a cabinet minister.

Green parties have also played an important role in government coalitions in a number of places and have influenced policy outcomes. To give a few European examples, in Germany, the Green Party was part of a coalition that governed between 1998 and 2005; in France, the Green Party contributed to the governing coalition between 1997 and 2002; and in Finland, from 1995 to 2002, the Green League was not only part of a coalition government but served in positions in the first and second cabinets of Prime Minister Paavo Lipponen. In Sweden the Greens are, at present, in a minority coalition government that began in 2014 and was renewed in 2019.

Outside of Europe, green parties have also been increasing in influence. In Australia for the past fifteen years Green Party candidates have received more third-party votes than any other non-dominant party and have held seats in parliament. The 2004 Nobel Peace Prize winner, Wangari Maathai, was the leader of the Green Party in Kenya and served both in parliament and as assistant minister for environment and natural resources from 2003 to 2005.

In other places, green parties play spoiling roles, preventing candidates from winning an absolute majority in the first round of elections or occasionally taking votes from other left-leaning dominant parties in a way that allows less environmentally concerned parties to prevail in elections. This phenomenon demonstrates the importance of understanding the structure of electoral rules, since the likelihood of electoral success of green parties depends on the political structures in which they are competing.

Green parties can also be important at sub-national levels – in local or regional governments, including in places such as the United States that don't have parliamentary systems. And they may also play a role in supranational governance – the first green party members in the European Parliament were elected in 1984, from Germany, the Netherlands, and Belgium; many more have served in that role since then.

Bureaucracies

In most political systems laws that are passed (especially at the national level) are fairly general, and the specific rules to implement them are made by agencies or departments of the government. These long-term structures of government, the ministries, offices, and other bureaucracies, remain in place even when the governing coalitions change around them. Most governments have one or more ministries or departments dedicated to addressing environmental issues. There may be a central one, such as a Ministry of the Environment or an Environmental Protection Agency. Other agencies or ministries, for example those pertaining to oceans, agriculture, or water, may focus on issues with strong environmental components.

One important aspect of this bureaucracy that oversees environmental protection is that much of it remains in place even when the political parties that run the government change. The actual heads of ministries or bureaus are likely to be politically appointed by whatever party runs the executive branch of government. But the vast majority of those who work in these offices are long-term staffers who remain regardless of who is in power. Although the priorities of government pertaining to the environment may shift (sometimes dramatically) with changes in elected officials, laws that are in place are much slower to be amended, and the continuation of personnel within bureaucracies reflects the continued trajectory of previous governments. The extent of personnel change within bureaucracies varies across countries and is an important factor to keep in mind when thinking about the functioning of governments.

The fragmentation of those departments can itself become a problem and contribute to the extent of governmental capacity to implement or enforce environmental policy (discussed in chapter 3). It can also lead to some of the consistency despite political change. One situation is what is referred to as a principal–agent problem, in which some entity (the "agent") is in charge of implementing action on behalf of a principal actor. Inevitably the agent's actions diverge from what the principal would have undertaken, in part because of incomplete information about the principal's intentions. These deviations usually coincide with the interests of the agent. Those who work in bureaucracies, for instance, want to continue their efforts and may therefore act in ways that further institutionalize their roles. One early demonstration of this phenomenon was in the context of the Environmental Protection Agency (EPA) in the United States during the anti-environmental presidency of Ronald Reagan; despite clear executive branch preferences, monitoring and enforcement of environmental rules by the EPA did not decline.[4]

oops, I always forget to click on the link

The Judiciary

Courts don't make policy, but they do ensure that rules are upheld. They help interpret these either by resolving aspects that are unclear when they are applied or by adjudicating conflicts over how to implement them. Courts are places where the specifics of those rules can be challenged when rules are created by incorrect processes (or are not created at all), deadlines aren't met, or other aspects of the law are not upheld. The role of this function can be especially relevant for contentious environmental politics; the side that did not prevail in the initial legislation can try to use procedure to get its way, and so political battles continue to be fought through the judicial process.

Depending on who within a political process is considered to have "standing" (the ability to bring a case within the court system), courts can also be broadly used by citizens or environmental organizations to compel the enforcement of existing rules. States such as the Philippines offer a wide acceptance of standing as a method of countering executive power, and some states, for example Australia and Italy, give standing to environmental organizations to compel the enforcement of environmental laws.[5]

Courts are also important because they are one of few places in political systems where penalties can be levied. The mere existence of a rule does not necessarily mean that it will be followed, and the costs of its lack of implementation fall to those who suffer from the environmental problem the rule was created to protect. This role of courts has been influential in cases of environmental injustice.[6]

Courts play particularly important political roles in governmental systems that are not unitary in one dimension or another. In federal systems they can help adjudicate differences between rules across levels of government, and in separation-of-powers systems they can resolve differences across the executive and legislative branches of government. In presidential and/or federal states such the United States, courts can take up challenges from one branch or level of government to another. Courts can also be used to force reluctant governmental officials to implement or enforce existing rules. For example, the state of Massachusetts sued the EPA during George W. Bush's administration and the case made it to the Supreme Court in 2007; Massachusetts argued – successfully – that the Clean Air Act requires the EPA to regulate greenhouse gases.[7] This ruling compelled executive branch action against the wishes of the administration. More recently, several US states have sued the EPA for rescinding climate change regulations issued by the previous presidential administration without going through a required review process.[8]

While all political systems have some kind of judiciary, this branch of government can vary in ways that have implications for environmental politics. Most important is how independent it is. Independence can be achieved in different ways: are judges elected or appointed, how long do they serve, and are their salaries guaranteed? Appointed judges with lifetime terms and guaranteed salaries are seen as the most independent, because they are not subject to political pressures in order to keep their jobs. Industry is more likely to be well funded and well organized than are environmentalists, so environmental interests may be at a disadvantage in working to elect judges that are sympathetic to their concerns. A judicial branch that isn't subject to those electoral pressures may be more likely to take environmental concerns into effect.

Independence in a judiciary can have downsides for the environment, though; a judicial branch that is insulated from political pressure is less likely to side with social pressure for environmental protection.[9] It's also clear that judicial action, especially coming into play at the end of political processes, is only one of many aspects influencing environmental outcomes, and the variation in environmental implementation across states cannot be accounted for by type or independence of judicial system.

Also important is the presence in some political systems of specialized environmental courts. Scholars such as Patricia Wald who have studied existing environmental courts find that, in general, they are beneficial to environmental outcomes, though others argue that they may not be necessary and that courts generally have a good track record of protecting environmental interests.[10]

Courts are reactive: they respond to perceived injustices after they have happened and thus are rarely able to be used to prevent environmental problems. That factor is baked in to the concept of "standing"; usually the only entities that can challenge the legality of something in court are the actual entities who have suffered because of an action that has taken place. That means both that only a specific set of actors can bring a case before a court and that they cannot do so in anticipation of a harm. In that way, courts cannot prevent environmental problems. But they can provide for compensation, which is otherwise rare in political processes. And most legal systems function on the idea of precedent: one legal decision then guides future decisions on that topic. So, when environmental cases are won, they can have a broader impact than just on the narrow case that was initially being decided.

Environmental Interest Groups

One way in which the public's concern about the environment is presented within the political process is through interest groups. These are organized efforts by citizens to put pressure on governments – or others – to accomplish various ends. Environmental interest groups have existed for many decades in many democracies and even in some authoritarian states.

Environmental organizations can play an important political role because they are able to prioritize the long term in a way that politicians, focused on the next election, find more difficult. As the scholar Karen Litfin points out, "most environmental problems will outlast the policymakers charged with addressing them,"[11] and environmental organizations have the ability to pressure politicians and give them a reason to focus on these issues.

Often these interest groups take the form of non-governmental organizations (NGOs), formally organized entities that are not a part of governmental structures. These civil society groups can play a variety of roles within political processes to advance environmental agendas. They can directly lobby government officials in the service of environmental outcomes, attempting to create regulations to protect the environment. Or they may work more indirectly, as "think tanks" or organizations doing research used to advocate particular positions or outcomes. Even organizations that are not directly trying to influence policy have a political role to play. Raising awareness of environmental problems can help create support for political action to address them. Influencing the environmental behavior of citizens, an important role played by less obviously political NGOs, may not only help address an environmental problem on its own but also make later policy options more viable, because they remove the opposition of some people who may already have adopted the desired behavior.

There are also "direct action" organizations that bypass the political process and try to stop problematic behavior more directly. Greenpeace, for example, has been known for such actions as putting activists literally between whalers' harpoons and the whales at which those harpoons are aimed. Although actions of this sort may temporarily prevent an individual whale from being killed, the strategy is broader than that. In part, activists put themselves in danger to demonstrate how strongly they feel that their cause is both important and just. The success of these activities comes in part from "bearing witness" to the problematic behavior – and, more importantly, gaining the sort of publicity that allows the public to see the actions that are being undertaken. Greenpeace has a sophisticated media operation and organizes its actions in ways that maximize exposure. The broader effort, then, is to influence the politics of this issue, albeit indirectly.

Other organizations take direct action precisely because efforts within traditional political processes have failed. Environmental justice movements frequently rely on direct action because they are disadvantaged in their access to political decision-making. Recent protests of this sort include, among many others, efforts (temporarily successful until the change of political administration within the United States) to prevent the Keystone XL pipeline from disrupting Native American land, action against lead in drinking water in African American communities in Flint, Michigan, protests in Chile and Argentina against ecologically disruptive gold mining practices by multinational corporations, and efforts to stop oil drilling that affected Ogoni people in Nigeria.

While most direct action environmental organizations begin from a principle of non-violence, there are even some activist groups (sometimes referred to as "eco-terrorists" by those who are opposed to their operations) who take action that is more clearly illegal or intended to cause harm or danger to entities that are causing environmental damage. In the United States, early forest-protection activists advocated – and practiced – "tree spiking," pounding metal rods into trees that were in danger of being cut down, so that the spikes would damage the equipment being used to cut down the trees or process them into lumber. (As with Greenpeace's anti-whaling action, the goal was less likely to cause the dangerous outcome – harpooned protesters or damaged equipment/injured loggers – than to persuade the whalers or loggers not to bear the risk their activity would bring.) Animal rights activists who release captive animals, or anti-development organizations that burn down corporate offices or ski lodges, also practice this form of politics.

Most discussion of the role of environmental organizations concerns domestic politics. Their kind of lobbying or political persuasion has the most access to organizations within the states (or even sub-state governments) within which they operate. But environmental NGOs also work to influence action in states other than their own, and on the international level (discussed further in chapter 5). They can, most obviously, advocate that their own state take positions internationally or work to address international issues that will have benefits outside of their state. Some environmental NGOs also work directly across borders to influence the behavior or politics in another state. There are more and more NGOs that are themselves multinational – membership organizations that work simultaneously in multiple states.

NGOs also play roles in international environmental negotiations. Many – at this point even most – multinational negotiating processes and organizations allow NGO observers, and some of these even allow NGOs to intervene in the formal proceedings to ask questions or make observations. In the negotiation of the Basel Convention on the Transboundary Movement of Hazardous Wastes and their Disposal (1989), two NGOs, Greenpeace and the Center for Science and the Environment, drafted language that ended up in the agreement.[12] NGOs have served in the delegations of states at meetings of the International Whaling Commission and have prepared briefing materials (and even paid membership fees and travel expenses for the delegations) for some states in order to support anti-whaling positions.[13]

These organizations play important informational roles in both directions: they can be called upon to give their expertise about potential solutions to environmental problems (or the costs or benefits of various proposed courses of action), and they also serve as a conduit of information to people outside the negotiating process, letting interested observers know what is happening in the negotiations (and sometimes what positions their own states are taking). One organization that has been especially influential in this way is the International Institute for Sustainable Development (IISD), which publishes the *Earth Negotiations Bulletin*, a daily update on negotiations and meetings of international environmental organizations as they are taking place.[14]

NGOs can work internationally in ways that circumvent traditional political processes to protect the environment directly. One such organization, Conservation International, buys plots of land, primarily in developing countries, to protect land directly, rather than relying on domestic or international political action to protect ecosystems.[15] Another early NGO action of this sort came when the concept of debt-for-nature swaps was popular. In these endeavors, during the debt crisis of the 1980s actors could "buy" the debt of countries on a secondary market. Any entity with the funding could do that, and what it meant is that the state would owe the money to that new entity instead of to the original lender. Because there were questions about the extent to which states would be able to repay, buying the debt cost considerably less than the original loan would bring in. What NGOs (and some countries) did was to buy the debt of highly indebted countries with a lot of biodiversity and essentially forgive the loans if the country promised to protect some amount of its biodiversity.[16] This concept was first proposed by Thomas Lovejoy of the World Wildlife Fund in 1984, and thus, no matter who bought the debt, it was the conceptual innovation of an environmental NGO.

Interest groups, whatever their approach, also need to take steps to overcome collective action problems (discussed in chapter 1). Even if many people in a community are concerned about environmental problems or would benefit from their resolution, the collective action needed to pressure the government to act may not be forthcoming. Free-riding – letting others do the organizing work – makes conceptual sense. After all, if environmental protest or lobbying manages to create policy that makes the air cleaner, the air is cleaner for everyone, not just the people who participated in creating the political pressure for new policy. And people have many concerns and busy lives – the environment is just one of many issues they may care about.

Environmental interest groups thus face a complicated process of persuading the public to participate in actions undertaken to protect the environment. They make use of different approaches to help overcome collective action problems. One is offering selective benefits – things that people get by joining or contributing to these organizations to which they would not otherwise have access. People who join the Sierra Club in the United States, for instance, may gain access to recreational activities the group conducts.[17] Some of these benefits may be "psychic"; the process of joining with like-minded people in support of something you care about can itself be a benefit of this type of political action. Similarly, the signaling element of joining an organization (and perhaps wearing its T-shirt or carrying its tote-bag) can be important socially and encourage people to participate in collective action.

At the same time, there may also be interest groups working against environmental action. Frequently they are strongly connected to industry, discussed further below, but there may be independent organizations that have other concerns, such as safety, that may have contrary approaches to an issue with environmental implications. Or there may be organizations advocating that funding or regulatory efforts be put to priorities other than environmental concerns – health, education, disability access, and other reasonable priorities. On any given issue it is worth examining what interest groups may be engaged by it and what their size, resources, and level of political access is likely to be. Environmental organizations also face the conundrum of simultaneously persuading potential members that their support is needed ("we can't do it without you!") and that the organization already has enough support and sufficient resources to accomplish its goals, so participating will not be in vain.

Business and Industry

Business and industry, as the underlying creators of many environmental problems, have a special role to play in environmental politics. Businesses do not intend to create environmental problems, but the nature of environmental problems as externalities to economic activity means that such problems may result simply by their engaging in their businesses. The action of seeking profit means minimizing costs. Since environmental problems come from externalities – which by definition means that the businesses that produce them don't bear a cost – simply working to earn a profit can create environmental harm. At the same time, because so much of what causes environmental problems comes, in some way, from business and industry, the actions of these groups can be especially important in resolving environmental problems. The politics of how environmental problems get addressed – or not – depends a lot on the interests and role of business.

Because externalities are unpriced (meaning that if, for example, a factory causes air pollution in its manufacturing process, it does not bear a cost from creating that pollution unless one is externally imposed), harming the environment is not generally bad for business. In fact, taking action to protect the environment can be costly. Businesses are generally working to minimize costs, so if there were non-costly ways to manufacture without producing pollution the factory operators would be equally happy to do so. The fact that the pollution is being produced means that doing so probably costs less than manufacturing in a non-polluting way.

In the long run, and collectively, making businesses operate in ways that do not harm the environment may not be inherently costly to society as a whole, and maybe not even to that business. But in the short run, and individually, it almost always is. If your business produces air pollution as an externality of its manufacturing process, mandating that it stop producing that pollution will be costly. Doing so will require expensive equipment, a change in production processes, or simply producing less efficiently, all of which cost money. For that reason, businesses are rarely in favor of rules requiring that they protect the environment (or rules that make it more costly to pollute or use resources).

Business tends to be influential in politics. There are structural reasons for this influence. Politicians are concerned about the economic wellbeing of their jurisdictions, and business plays an important role in the economic health of countries. A business employs people, sells things that others want, and brings money into the area. For that reason alone, business has a powerful role in the political process.[18] Fears that environmental rules that make manufacturing or selling things more expensive and will thus lead to less profit and less economic activity can be played up by businesses opposing environmental rules. These businesses can threaten to move to other areas with lower environmental standards, taking their economic benefits with them. Businesses also tend to have more money than concerned citizens to put towards achieving their political goals.

Part of this influence of business can be traced back to the role of environmental problems as collective action problems, as discussed in chapter 1. While there are potential collective action problems on all sides of environmental problems, it is important to note how many actors there are and what their interests are. Environmental problems tend to involve diffuse interests. That characterization has two elements. First, environmental problems are often experienced by large numbers of people. You might expect that to be a political advantage – many people in favor of addressing a problem. But that also means many people who would need to cooperate successfully (overcoming collective action problems) to work effectively for change. Generally, the more people who need to engage in collective action to make it happen, the more difficult it is – each person contributes only a tiny fraction of the effort and knows there are so many other people who might also be interested in addressing the problem. Collective action problems with large numbers of people can be especially difficult to overcome.

A second way that environmental problems are diffuse is that they are likely only one of many concerns of those who are experiencing them. Someone worrying about air pollution from a local factory may also be concerned about healthcare, getting a good education for her children, and ensuring that she keep her job, among myriad other interests. Given her limited time and resources, she may not be able to contribute to addressing all problems, and so the anxiety about the environment, even if it exists, may remain unexpressed or not be acted upon.

Business is on the other side of this equation, with interests that are concentrated. The factories worried that anti-pollution rules would harm their economic profitability are limited in number and thus more likely to be able to organize to oppose political action. At the same time, the number of factors they are concerned about is likely to be smaller, and avoiding rules about pollution is likely to be high on their agenda. There are good reasons for that: changing behavior to avoid producing externalities is almost certain to be more costly, in the short term, than would otherwise be the case.

The cost of production of substitutes for environmentally harmful substances will decrease over time if environmental policy requires or incentivizes their use. Innovation will likely contribute, as will economies of scale: as more of something is produced, the per unit cost of making it generally goes down. That happens for many reasons, including the creation of big fancy new factories that can produce these substitutes in large quantities and thus require less equipment or personnel per amount produced.

Sometimes businesses find alternative ways of producing when they are required to make changes to protect the environment. The "Porter hypothesis" suggests that businesses and other organizations aren't as efficient as they could be because of incremental development and standard operating procedures. There might be cheaper (and environmentally friendlier) ways to do things that they don't consider until regulation or price changes force them to go looking. So, in some cases, business costs may go down because of environmental regulation.[19] The field of industrial ecology builds on this insight, focusing on things such as using waste heat from one process to provide electricity for another and finding wastes in one business that can be used instead of raw materials for another. The 3M company in the United States discovered that it could save more than $4 million annually by making reusable packaging for shipping, and thereby both reducing the costs of buying new packaging and the cost of disposing of waste from the old ones. A Mexican company that makes car engine manifolds, Industrias Fronterizas, figured out that separating waste products from clean water via a filtration unit saved the company more than $1 million annually by reducing both machine downtime and overall waste.[20]

Regulations can underpin the advantages identified by industrial ecology. If it costs more for businesses to dispose of waste, they are more inclined to reduce it or find other businesses for which their waste products could be productive inputs. Substitutes used instead of environmentally damaging chemicals may become cost effective when the harmful substances they replace become more expensive. As chapter 1 explains, non-renewable natural resources (such as oil) or even some renewable ones (such as fish) provide that incentive on their own; as they become depleted they become more costly, and it pays to develop ways to use them more efficiently or to find substitutes. Regulations can create that same incentive. "Most companies will only be as green as governments make them," as scholar Frances Cairncross notes,[21] whether the government action comes from direct rules or from creating incentives that make it more costly to continue to pollute.

For all of these reasons, regulating businesses or environmental activity as a way of addressing environmental problems rarely turns out to be as difficult or expensive as it is anticipated to be. And, obviously, the aggregate social costs of causing polluting activity to become more expensive for businesses can make collective economic sense, because the costs to society from environmental damage decrease when that damage decreases.

That being said, even when it makes overall long-term economic sense to address environmental problems, the costs and benefits are not distributed equally. Those (such as businesses or industries) that are contributing to the environmental problem may indeed be made worse off by having to act to prevent it. That leads them to participate in political processes to try to prevent such regulation and often to have an outsize effect in their efforts.

Some political structures are also set up in ways that are especially friendly to business interests. Corporatism describes political structures in which corporations exert control over the governing apparatus of the state. The purest example of a corporatist state was Italy during its fascist period between the two world wars; in the modern era, various Scandinavian states such as Sweden are considered neo-corporatist. Even apart from strictly corporatist states, there are states in which there is a close relationship between government and industry, with politicians coming from the business world and – perhaps more importantly – going back into important roles in corporations after serving in government. This relationship grants businesses special access to, or knowledge about, those in government and may thus help reflect the interests of those businesses in the political process.

Business can also play important roles in protecting the environment, including working for environmental ends within – or at least not in opposition to – political processes. As the secretary general of the United Nations Conference on Environment and Development (UNCED) put it, "The environment is not going to be saved by environmentalists. Environmentalists do not hold the levers of economic power."[22]

There are businesses that benefit from environmental action, and are thus more likely to support it, including those that manufacture alternative substitutes, processes, or technologies used to address environmental problems. Not surprisingly, their interventions into politics are likely to be on the side of favoring action that would increase the odds that people or other businesses would use whatever it is they are manufacturing or selling. Manufacturers of solar panels, for instance, may benefit from rules requiring or incentivizing use of alternative energy.

There may be other reasons the businesses will not necessarily oppose environmental action. They may recognize that their future costs – or, in particular, future liability – may not be as high if they produce in more environmentally friendly ways. They may also benefit from a "green" image in terms of garnering investors, employees, or customers. In the latter case, consumers interested in the environment may be willing to pay a premium for environmentally friendly or sustainably produced goods. In those situations, even if manufacturing or operating in an environmentally benign way costs more, the additional premium their products can fetch by being labeled as "green" might make up for that cost.

If everyone is mandated to produce in a sustainable way, the benefit to those who do disappears. But, at least initially, those who already do may be able to outcompete those who have to change behavior to meet new mandates. For that reason alone they may support mandates for environmentally friendly behavior (or may, in anticipation of these mandates, work to improve their environmental record before being required to do so). Businesses, fearing regulation, can also promise to behave environmentally (or even do so) in an effort to suggest that rules are not necessary. There may also be advantages to businesses in accepting local regulations, where they can find compromises with people whose economic lives are tied to the industry, before regulation at national or international level becomes more likely. If they operate in a legal system that holds actors responsible for accidents or other environmental problems, businesses may further choose to address environmental issues even when not required to, because of fear of liability.[23]

There is plenty of what is known as "greenwashing" by businesses – efforts to paint actions that benefit business, or have little benefit to the environment, as environmentally friendly. Businesses do this to take advantage of consumer interests in doing the right environmental thing – presenting buying their products as a solution to environmental problems. The larger critique of green businesses is that, if their underlying products or processes are environmentally harmful, producing a greener version (say, of disposable dishware) only allows the environmentally problematic actions to continue.

Although this type of business behavior may not appear relevant to the political process, its potential for draining citizen pressure for action by channeling it into consumer behavior can have important political implications. If people are lulled into believing that they're doing their part for the environment by buying greener products, they may be less likely to engage in political action or support policies that require more significant environmental change.

Business–Environment Coalitions

Environmental action may become politically possible in the presence of what some have referred to (in an analogy to Prohibition in the United States) as coalitions between "Baptists and bootleggers" or "the green and the greedy," referring to nominal opponents who find common ground. In other words, it is when there is some overlap between the interests of environmentalists (or environmental organizations) and business or industry that environmental solutions may be possible.

That can come when businesses seek protection from competition and environmental organizations seek environmental improvement. For example, when the United States had costly rules that protected sea turtles from shrimp fishing and other countries didn't, the first choice of US shrimpers was to try to get those rules removed. But their second choice was to ensure that shrimpers with whom they competed for markets were held to the same costly rules. At the same time, environmental organizations knew that sea turtles would be adequately protected only if shrimping vessels from other countries also protected them. Each set of actors was lobbying lawmakers for its preferred outcome. Not surprisingly, the resulting policy restricted the import of shrimp that was caught in ways that didn't meet US standards, the policy outcome that represented the overlap among the two groups' goals.[24]

Air pollution policies that apply only to new power plants or factories but that allow existing plants to continue using old technology might be the kind of compromise between what environmentalists want and what industry wants that can gain acceptance within a political process. Regulation that results from this kind of convergence of interests would be likely to focus on disadvantaging new entrants to an industry rather than existing industries. Looking at where the interests of businesses and environmentalists overlap can thus be either a useful tool for creating political action or a predictor for when such political action is likely to emerge and what shape it may take.

Media

The media – newspapers, television, the internet, among others – provides information that the general public uses to form opinions about many topics, including the environment. These outlets thus become a site of contention: not neutral presenters of apolitical information but, rather, being used by the various actors engaged in political struggle to attempt to influence public attention. Some characteristics of both old and new media contribute to their contentious and problematic nature with respect to environmental information.

In many countries, including the United States, there is an ideal of journalistic objectivity. This approach suggests that, when there are controversial matters, "both sides" of the issue should be presented. But that approach privileges minority opinions which can argue that there is disagreement on a topic and that "their side" deserves as much attention as the dominant position. Because journalists are not likely to be much more scientifically literate than the general population, and may be working on tight deadlines and on stories outside their area of expertise, they may not have the ability or inclination to judge how valid are the various potential "sides" of an issue. Those who oppose environmental regulation can take advantage of that framework to manufacture disagreement by presenting information – often generated with funding from industries that are complicit in creating the environmental harm[25] – about whether an environmental problem is real. Journalists may also be amenable to this framing because a controversy creates a more interesting story than an uncontested piece of information.

A random sample of peer-reviewed scientific publications about climate change showed that not a single one questioned the scientific consensus that climate change was happening and was caused, at least in part, by people;[26] at the same time, a study of newspaper stories of climate change found roughly half included equal attention to "both sides" of the question as to whether climate change is happening and humans contribute to it, and another third supported the idea that climate change was real but nevertheless included dissenting details.[27] While in recent years the accuracy of journalistic coverage of climate change has improved, the propensity towards including contrary information persists.

Social media, user-generated content without the formal editorial oversight of traditional media, continue some of these circumstances but upend others. The primacy of controversy is augmented, because this kind of content is premised on gaining attention and being shared, and presenting disagreements or contrary points of view gains attention. Social media also augments confirmation bias, by giving people the ability to tailor their media environment so that they encounter only the kind of information they prefer. Perhaps more concerning is the rise enabled by social media of false stories, planted either by citizens pursuing a political agenda or, as the 2016 US election demonstrated, by foreign governments attempting to undermine political order. This misinformation sits uneasily with our cognitive biases about risk and our psychological defense mechanisms that prefer not to consider worrisome information that might require difficult choices.

Conclusion

We need to know the interests and capabilities of the various entities who engage in political processes – be they environmental organizations, businesses, or politicians – as a starting point for understanding environmental politics. These actors engage with the political process to advocate for their interests, and the more access they have, and the stronger their support, the greater opportunity they have to influence political outcomes.

At the same time, other actors mediate or direct those interests within the constraint of the political structures discussed in chapter 3. Political parties represent the interests of those who favor or oppose environmental action, and how many dominant parties and how focused they are on the environment may determine whether anyone serving in government begins with a concern for environmental issues. A state's judicial processes provide another avenue for battles over what rules require or whether they are being adequately implemented. The media, both traditional journalism and newer social media, shapes public opinion by how environmental issues are covered. In practice, the norms of traditional journalism, and the freewheeling nature of social media, have lent themselves to capture by industry-related organizations trying to persuade the political process not to act on environmental issues.

Understanding both the interests of the political actors and the contexts and methods they use to pursue them is central to understanding, or predicting, political choices about whether or how to address environmental problems.

CHAPTER 5

International Environmental Politics

Environmental politics at the international level is different from local or national environmental politics in important ways. At the domestic level, governments make rules that bind their citizens. Though those citizens may play a role in advocating for or against those rules, they can be held accountable for upholding them, even if they would prefer not to. At the international level there is no overarching authority that can create rules for states (which is the term that refers to what is colloquially known as "countries") or require that they be followed. Any rules that bind states internationally must be made by those states. That dramatically changes the extent and type of international policy and the politics of addressing environmental problems internationally.

The complexity of international negotiation includes a context in which negotiators are concerned with both their domestic populations and the other states with which they are negotiating. States negotiating internationally also face a trade-off between stricter rules that bind fewer states (those willing to take on strong regulation) and weaker rules to which more states agree. Once any international agreements are reached, implementation, compliance, and enforcement are also more difficult than in the domestic context.

As a result of the process that creates them, international rules are fragmented, diffuse, and overlapping, involving the creation of multiple treaties and institutions to manage them, and are often weaker than domestic rules. There are hundreds of international agreements addressing environmental problems. In part because of the complexity of creating binding international rules, other forms of international governance help address environmental issues. Non-state actors may work across borders to influence the behavior of actors in non-regulatory ways.

The interaction between domestic politics and international governance plays an important role for both levels of environmental politics. This chapter concludes with an examination of the effect of actions taken at the domestic level on international environmental politics, as well as the effects of international cooperation, or the lack of it, on how these issues are addressed within domestic political contexts.

The Framework for International Environmental Politics

It makes sense that we would work to address some environmental issues on the international level. Many environmental problems cross borders in some way. Some types of air pollution, such as sulfur dioxide that can become acid rain, may be created in one country and move with airflow to another country, where the problem is felt by those who played no role in creating it. In that situation, nothing the receiving country does on its own can prevent its population from experiencing the pollution. Global climate change is even more diffuse. The greenhouse gas emissions that cause it occur largely within states, but the problem itself is global: where the effects are felt has almost no relationship to where the emissions take place. In other cases, resources themselves move – for instance, migratory species. In these cases, one country can protect birds or the wetlands on which they rely (for example), but if the country to which the birds migrate does not protect them or their habitat, the species will decline nevertheless.

Other resources are inherently international. No state has individual jurisdiction over the high seas. Because of that, countries have overharvested fish from, and dumped pollution into, the oceans. The atmosphere is likewise outside the control of any one state, the ozone layer, located 10 to 50 km above the earth, cannot practically be regulated by any one state, but what happens to it can affect all of them.

The starting point for understanding international environmental politics is what those who study international relations term "anarchy." Rather than meaning chaos, the formal definition of anarchy is that there is no actor with overarching authority. In other words, there is no one, be it another state or an international organization, that can tell a state what it must or cannot do. Anarchy is the centerpiece of international relations, and it means that, in order for states to be bound by a rule, they must collectively agree to it.

International law, therefore, is something into which states voluntarily enter. That doesn't mean it's not legally binding once they have taken it on. But, unlike domestic politics in which a person who doesn't consent to a speed limit (or a requirement not to put pollution into a river) is still required to follow that rule, states that decide not to participate in an international agreement cannot be bound by its rules. That makes international agreements a particularly difficult form of collective action problem.

As chapter 1 discusses, environmental problems in general are a form of collective action problem and also common pool resource problems. States that remain outside of an agreement to protect fisheries and continue to fish, for example, can thus prevent others from being able to protect fish stocks. Within states (or other sub-state jurisdictions) this problem can be overcome – though not always easily – by action taken by governments. Domestic political processes can make rules that tell their citizens that they must not put toxic materials in water or can catch only a certain amount of fish per year. That kind of rule making can't be done at the international level, and that affects the process and nature of international rules.

International rules are generally weaker than domestic rules because of the need to ensure that all relevant states participate. Any state that threatens to stay out of an agreement (usually because it would prefer to keep engaging in the activity that the agreement will prohibit) gains a lot of bargaining power. Other states need it to join for the agreement to succeed, so it can mandate that rules be made weaker or that it get some other form of compensation before it is willing to take them on. International environmental agreements often succumb to what is called "least-common-denominator" bargaining – in other words, rules are set at the level that the most reluctant state will accept. They are therefore weaker than at least some states would prefer, and often weaker than environmental conditions – and the scientists who study them – would suggest.

→ pretty important

States face what scholars refer to as a "two-level game" when engaging in international negotiations.[1] They are negotiating with other states concerning the rules and processes but at the same time also with their domestic population watching the proceedings. On the one hand, they are genuinely constrained by what their domestic populations want and can't agree to international rules their populations aren't willing to go along with. On the other hand, state negotiators make use of their domestic population as a negotiating strategy. One primary approach is to use the domestic population as an excuse for being unwilling to accept more stringent rules. The negotiators can indicate that they, the negotiators, might think the

→ still need
between something
to implement policies

Authoritarianism form of GW

proposed option is a good one but that their domestic political process will never accept it, and they are therefore restricted as to what they can accept. They are also constrained in their ability to accomplish what their domestic populations want but other states won't agree to.

Acceptance by the domestic population of the results of international negotiation is important because of how international law actually works: states agree to a rule, but it then requires "ratification," which involves some type of acceptance by the state's political decision-making process. That process varies by country: for the United States, for instance, treaties need to be ratified by two-thirds of the Senate; in the United Kingdom, Parliament has twenty-one days to approve a treaty after a prime minister signs it; in Japan both houses of the Diet (the national legislature) must approve a treaty before it becomes legally binding.

That ratification process is not always forthcoming, particularly in separation-of-powers states (discussed further in chapter 3). The United States illustrates this difficulty. A number of international agreements that US negotiators have signed have failed to be ratified. For example, in 1997, while the negotiation of the Kyoto Protocol to the United Nations Framework Convention on Climate Change was ongoing, the US Senate actually passed a resolution indicating that it would refuse to ratify an agreement that didn't contain emissions reductions obligations for developing countries,[2] which was at that point off the table.

Likewise, the United States has signed but not ratified the United Nations Convention on the Law of the Sea (1982), the Basel Convention on the Control of Transboundary Movements of Hazardous Wastes and Their Disposal (1989) and the Convention on Biological Diversity (1992), among others. Because the United States does not have a parliamen-

such a
divide

tary system, control of the executive branch (which negotiates agreements) and the Senate – one house of the legislature (which ratifies them) – may be held by different parties, making ratification more difficult than in parliamentary systems.

It was largely because of this structural impediment from the United States that the most recent international climate change treaty, the Paris Agreement on Climate Change (2015), was negotiated in the form that it was. Rather than negotiate a traditional agreement in which countries committed to specific levels of emissions reductions, the Paris Agreement requires that states set "nationally determined contributions" to emissions reductions; these can take any format and include any type or level of commitment. The Obama administration argued that, because states were not required to undertake any specific formal international commitment, the agreement could be signed by executive order and thus did not require Senate ratification, which would not have been forthcoming.[3] President Obama and the other negotiating countries knew that the Senate would never approve it, and thus from the beginning the agreement was negotiated in such a way to allow this action by the United States.

International agreements also contain provisions for countries to cease to be bound, though doing so usually requires a certain waiting period before the withdrawal takes effect. These provisions are included because states would be reluctant to agree to laws that would bind them in perpetuity. In recent years, Canada withdrew from the Kyoto Protocol when it became clear that it would be unable to live up to its obligations.[4] After years of threatening to do so, Japan has just withdrawn from the International Convention for the Regulation of Whaling (1946), because it doesn't like the commercial whaling moratorium the agreement's governing body has put in place and it has been unable to change the policy.[5]

The United States, under the Trump administration, has also threatened to withdraw from the Paris Agreement on Climate Change (2015) that was accepted by the executive branch. (The catch, however, is that the agreement stipulates that it must have been in force for three years before a member state can indicate its intention to withdraw, and even then the withdrawal becomes effective only one year after that – which, coincidentally, would be the day after the 2020 US presidential election.) So, despite popular perceptions, the United States has not – as of early 2020 at least – withdrawn from the agreement.[6]

Even when states agree to be bound, enforcement can be difficult – it's what scholars refer to as a "second order problem."[7] Any enforcement needs to be undertaken by the group of states that have agreed to a regulation, because there's no external authority in international relations that can make anything happen. Enforcement, therefore, also poses a collective action problem. States need to agree on what will be done to detect and punish any violation of the rules, and they also need to step up and collectively do it when enforcement is needed. In the same way that states may be tempted to free-ride on the rules (either by not joining or by breaking international agreements), they may be tempted to free-ride on the monitoring and enforcement provisions of these agreements.

These factors combine to make agreements weak and enforcement measures even weaker. States are unlikely to agree to strong penalties for violations of agreements, because they might themselves be subject to them. (And, as Canada's experience with the Kyoto Protocol suggests, if strong penalties for non-compliance are enacted, they might simply persuade non-compliant states to withdraw from the agreement.) As a result, most non-compliance procedures within international environmental law are designed either to shame states back into compliance, by exposing their wrongdoings to the world, or to assist those states in doing what they need to do to comply, often with technical or economic assistance.

However, we should not be overly concerned about the enforcement problem and the weakness of international compared to domestic politics. Even though there are enforcement mechanisms in domestic law, they are often not the primary reason that individual or corporate actors do the things they are bound by law to do in those contexts. Enforcement may be stronger in domestic contexts than in international contexts, but for any given action it is unlikely that misbehavior, should it happen, would be detected and punished. Laws, both domestic and international, work primarily because they are generally accepted by the actors within the system. (Remember, the agreements that get made in the first place have already managed successfully to overcome collective action problems to address environmental issues states could not tackle on their own.) Nevertheless, the lack of strong enforcement in international relations is a feature of the system and does cause problems in environmental politics.

The inability, within some types of agreements, to detect non-compliance is likely a greater problem than the lack of strong enforcement measures. As one important scholar of international law notes, "almost all nations observe almost all principles of international law and almost all of their obligations almost all of the time."[8] But non-compliance does happen with international environmental law. Often a state may fully intend to abide by an agreement, but sub-state actors – the ones whose behavior ultimately determines compliance – lack either the ability or desire to do so. Efforts to protect endangered species suffer from this type of non-compliance, and the fact that so many people need to avoid harming endangered species in order to protect them makes it quite easy for non-compliance to happen and hard to detect it when it does.[9] In other cases, though, states may direct non-compliance in areas where it is hard to uncover. Illegal whaling practices by the then Soviet Union were approved at the highest level and came to light only after the Soviet Union's political collapse.[10] Because whales are caught far out in the ocean, it can be difficult to determine if the self-reported catch numbers are correct. (In more recent years, in part because of the fear of this type of non-compliance, international fisheries agreements have included monitors from other states on fishing vessels.)

Another difficulty for international environmental politics arises when the elements of an international environmental problem involve one or more states as primary sources of a problem while the effects are felt by a different set of states. In domestic politics, decision-making by a national government can address a problem that has both winners and losers within the same political boundaries. At the international level it can be difficult for a state affected by an environmental problem for which it isn't responsible to persuade a state that creates that harm to change course. Imagine a river that flows from one country to another. If the first country takes most of the water from that river, it won't be available for the second country to use. But the first country is not negatively affected by its water use; in fact, it would be harmed if it stopped taking so much water, because it has domestic constituencies that rely on access to that water.

The same thing can happen with pollution problems. A state that dumps its pollution into a transboundary river may not be affected at all (and would bear a cost if it had to come up with a different way to dispose of its waste); the downstream state suffers the consequences and isn't able to do anything on its own to prevent the problem. These types of problems can be especially difficult to address internationally. In many other types of environmental problems, such as ocean pollution, most states may contribute to the problem but also experience some of its harms, which gives them more of an incentive to address it.

It is reassuring to discover that even these directional transboundary issues, such as acid rain (which moves directionally with wind currents) or questions of rivers, are often addressed at the international level. In these contexts it can help if the states in question have other international interactions on which they might want cooperative partners. This allows for "issue linkage," essentially trading off benefits where one state might get something it wants in one negotiation and the other state may get something it wants on a different issue.[11] In contexts in which states interact a lot, this may take the form of what is called "diffuse reciprocity" – in other words, states generally expect that, over time, this give and take on issues will work out to their advantage rather than keeping track of specific benefits they are trading off.

And, of course, the time-horizon problems that can plague domestic efforts to address environmental problems exist at the international level as well. Dealing with the mismatch in time between when environmental problems are caused and when they are felt (or fixed) is harder at the international level than within states, because there is less control over whether all the relevant actors will live up to their obligations and eventually protect the resource over the long run. For example, efforts to reduce ocean fishing in the present will result in increased stocks and more lucrative catches in the future. But if your success at addressing overfishing requires all states to catch fewer fish in the short term in order to rebuild stocks so that everyone can catch more fish in the long term, not being certain whether others – because they are in other countries over whose behavior you ultimately have no control – will live up to those obligations might make that sacrifice in the present less likely.

How International Environmental Cooperation Happens: Ozone Depletion as an Example

Despite the difficulties the international system poses for environmental politics across borders, states do make policy at the international level, and it can make a meaningful difference in environmental problems. How does it happen?

The short answer is that, if states suffer from international environmental problems, international cooperation is frequently the only tool for addressing them. Addressing collective action problems benefits those who experience them, and by definition these problems cannot be solved alone. So international environmental action can happen when there are problems that states benefit from addressing and believe that enough other states care about. International organizations, and other mechanisms to increase trust among states, can help make that cooperation more likely. Other ways in which the cooperative agreements are designed can also contribute.

The global effort to address ozone depletion provides a useful example. The ozone layer, a thin protective layer of the O_3 molecule approximate 10 to 50 km above the earth, protects the planet's ecosystems and inhabitants from ultraviolet radiation from the sun. It turns out – though no one knew it at the time they were invented – that a class of industrial chemicals used in refrigeration, fire suppression, and other important activities could themselves break down in the presence of sunlight and cause a chain reaction, destroying O_3 in the stratosphere.

As discussed in chapter 2, one essential element of the international effort to address ozone depletion was gaining a basic scientific understanding of the problem and its consequences. As with other recent international environmental problems, efforts to address ozone depletion began before it was clear that a problem even existed; scientists theorized the possibility that one would exist.

At the same time that scientists were studying the phenomenon, trying to determine whether these substances were indeed present in the stratosphere and whether they were leading to a decrease in the thickness of the ozone layer (and, thereby, causing problems on earth), politicians began negotiating an international agreement to address the potential problem. In 1985, within the context of the United Nations Environment Programme, they completed the Vienna Convention on the Protection of the Ozone Layer. This framework convention lays out the context for collaboration and the broader principle that states should take "appropriate" action to protect the ozone layer; it created a robust process for evaluating the science of ozone depletion but did not yet have specific obligations for changing emissions behavior.

This agreement was followed shortly thereafter by the negotiation of what became the Montreal Protocol on Substances that Deplete the Ozone Layer (in 1987). By this point, what came to be known as the "ozone hole" – a seasonal thinning of the ozone layer over Antarctica – had been discovered, and states were willing to commit to action. The initial agreement simply called for a freeze and then a gradual reduction of the consumption of five major types of ozone-depleting substances, but it also put into place a mechanism for changing those obligations over time. An effort was made to encourage broader participation, first, by requiring that states inside the agreement refuse to trade ozone-depleting substances with those outside of the agreement and, second, by allowing developing states an additional ten years before they would have to meet the various obligations.[12]

When those efforts failed to bring important developing states (notably China and India) into the agreement, member states negotiated an agreement that specified what came to be known as the Montreal Protocol Multilateral Fund, a financial mechanism by which the "incremental costs" of meeting the obligations of the agreement would be met for developing states. In other words, when developing states used substitute chemicals that cost more than the originals would have cost, any additional price for those substitutes would be paid by this fund. This mechanism brought reluctant states into the agreement. The action by China and India demonstrates the negotiating power of reluctant states. Because those states were needed in the agreement, which was not among their most important priorities, other states that wanted to solve the problem had to take steps to keep them from being free-riders and thereby undermining the agreement.[13]

The results of the international efforts to address ozone depletion are no less than astonishing. The Montreal Protocol was the first international agreement to achieve universal ratification – literally every country in the world signed and ratified it.[14] The agreement has been regularly amended to include additional substances and to move from what was initially simply a phase-down to a complete phase-out of ozone-depleting substances, as well as to move up the timetable for phase-out. For the most part, the world has transitioned away from using these substances, despite their usefulness in industrial processes.

Many ozone-depleting substances have extremely long atmospheric lifetimes – the amount of time they exist in the stratosphere and remain capable of contributing to ozone depletion. The most common types of CFCs can persist for

100 years or more, and some types can last as long as 1,700 years.[15] That means that recovering from nearly a century of use of these substances may take a long time. Nevertheless, their release has almost entirely stopped, and the stratosphere is responding, with the ozone layer improving by between 1 and 3 percent per year since 2000. As of 2018, the seasonal thinning over the Antarctic, which is most obvious evidence of depletion, was 16 percent smaller than it was at its peak. The United Nations body that oversees the Montreal Protocol predicts that it will heal entirely by 2060.[16]

The process of international cooperation to address this issue hasn't been entirely smooth. In the early years of the Montreal Protocol a black market in ozone-depleting substances emerged. In the developed world it cost less to use illegal ozone-depleting substances to recharge air conditioners (especially in cars) than to retrofit them to be able to use the newer, more expensive substitutes; those who could get hold of CFCs could benefit from using them. This black market was made possible, in part, by the time-lag that developing countries were allowed before having to meet the agreement's obligations. During the time when developed countries were phasing down – and then out – their use of regulated substances, developing countries were allowed to increase their use; that meant that these substances were still globally available and could make their way to places they shouldn't have been used. But this black market, as well as other issues of good-faith non-compliance (in which states were attempting to comply with the rules but fell short) were short-lived.

Compliance with the agreement overall has been high for several reasons. First, the states of the world really do benefit from preventing ozone depletion, which helps them overcome the collective action problem of international cooperation. Benefiting from collective action isn't necessarily enough to overcome its difficulties (since there are still often benefits from free-riding). But it helped that the countries most concerned about the problem were those that were the wealthiest and thus initially the most able to deal with the required changes. More importantly, the mechanisms pioneered by the agreement to bring in reluctant countries – which have since become key tools across other international environmental agreements – did make it easier for developing countries to switch over to processes that did not deplete the ozone layer without initially having to bear a major cost.

Also key in the successful transition was the technology change that resulted. Substitute chemicals – and processes that could accomplish the same goals without using either ozone-depleting substances or their substitutes – were invented because those who could come up with these alternatives could make money from them, since the world would have to cease using the problematic chemicals. Once everyone was put on notice that they would have to stop using ozone-depleting substances, businesses knew they could benefit from manufacturing substitutes, and industry was willing to buy them.

These substitutes were initially much more expensive than the ozone-depleting substances, but as more and more were invented, and as they began to be produced in larger quantities, the cost per unit decreased. Moreover, once new technologies were adopted, there was no longer any advantage to using the banned chemicals. This technology transition accounts in part for the fact that the black market ultimately disappeared. Once automobile air conditioners were no longer made in a way that could use ozone-depleting substances, and older cars with the older technology were no longer in use, there was little need for the banned chemicals. The same is true for other uses of these substances.[17]

Compliance hasn't been perfect, however. Recently scientists have detected an increase (or, more accurately, less of a decrease than the phase-out would predict) in CFC-11, one of the most common ozone-depleting substances before international cooperation to address ozone depletion. Evidence suggests that Chinese factories are most likely behind this non-compliance, though whether they account for the full increase is unclear.[18] While this non-compliance is unfortunate, the monitoring and scientific cooperation created by the agreement allowed it to be detected and publicized, and its effect on addressing the environmental problem is not likely to be large.

What's more impressive is that ozone depletion is precisely the kind of environmental problem that should be especially difficult to address internationally. It is a true problem of the global commons, in which any state that doesn't participate in the solution can undermine the ability of others to address the problem. (The ease of making ozone-depleting substances also means that it was within the technological reach of any country, so remaining outside of the agreement and continuing to deplete the ozone layer was a realistic possibility.) The substances that caused the problem were an important component of industrialization (and cheap, easy, and safe for users), so people were reluctant to change their behavior, especially in light of an environmental problem whose effects were not yet obvious. And yet the world managed to cooperate in a way that fundamentally changed a set of industrial processes internationally and largely addressed an environmental problem before it became irreversible.

Ozone depletion shares a number of characteristics with climate change, giving some people reasons for optimism that we might be able to address that especially pernicious environmental problem. Ozone depletion is nevertheless easier to address than climate change in a few ways. First, with small exceptions, most ozone-depleting substances are human-made rather than naturally occurring. In climate change, while actions by people are causing the bulk of the problem, many of the chemicals in question are part of natural processes. For ozone depletion, any increase in the use of these substances can be attributed to human activity; that is not the case for climate change. In addition, both the number and the scale of human activities causing climate change are much greater than those implicated in ozone depletion. And, although there is great variation across the types of effects of climate change, it is also the case that many of the countries that will suffer the most are developing countries, while richer countries are more likely to be able to adapt to some of these effects. That makes rich countries somewhat less likely to take the lead in addressing the issue, even though it is their behavior that has caused the bulk of the problem historically. Nevertheless, the impressive global action on ozone depletion does give us reason to hope that major problems of the global commons can be managed.

Characteristics of International Environmental Agreements

The nature of international environmental politics leads to a landscape of international environmental agreements with certain characteristics. Agreements are numerous and sometimes overlapping (or even contain contradictory obligations). They are often far weaker than actually demanded by the environmental problems they address. The politics of how they are created is essential to understanding their form as well as their shortcomings and successes.

There is no overarching authority in charge of creating or even managing international environmental law. Even the United Nations Environment Programme, the arm of the United Nations that attempts to coordinate responses to environmental problems, is to some extent reactive: it can convene negotiations, but these only happen when states have shown an interest in addressing an issue, and it is not the only venue in which international negotiations take place. Because states can decide to work together to address international issues whenever, and in whatever context, they choose, most international environmental rules are passed by creating an entirely new international agreement to address a given problem, even if some related organization or legal mechanism may already exist.

There are several different types of international environmental agreements, explained by the political contexts in which they are negotiated. A common model, as seen in the ozone-depletion agreements discussed above, is what is sometimes called the "convention-protocol" model. In this approach, a broad framework convention is first negotiated that lays out agreement in principle to addressing the environmental problem but does not contain substantive obligations to change behavior on the issue. The initial agreement does, however, set out processes for cooperation, including – importantly – scientific cooperation. Once there is general agreement on these details (and, often, once scientific cooperation leads to further information about the parameters and severity of the problem), protocols that address the actual abatement measure necessary to address the problem can be negotiated. As in the case of the ozone-depletion agreements, either multiple protocols or multiple amendments to protocols can be negotiated over time, allowing the agreement's obligations to change as more information becomes available on the severity of the threat or on new substances to regulate.

Despite their advantages, one of the primary reasons that framework conventions are initially negotiated is that there is insufficient agreement by states at the beginning of the process to be willing to agree to stringent abatement measures. The negotiations often begin with the hope of crafting a set of abatement obligations, but, when that is not possible, the states settle for agreeing on the principles of protecting the resource in question and a set of processes for continued research and later negotiation. This type of agreement has become more common in international environmental politics over time, in part because, as the remaining international environmental issues become more complex and we begin to try to address them before their full effects have appeared, politics becomes more contentious. It can be difficult to gain agreement to take current, potentially costly, action for uncertain future collective benefit. Another disadvantage of the convention-protocol model is that negotiating each new protocol or amendment takes time, and each has the legal weight of an entirely new obligation, requiring signatures and ratification by the participating states, a certain number of which need to ratify before the protocol or amendment becomes legally binding on ("enters into force" for) its participating states.

Many agreements – sometimes called comprehensive agreements – are nevertheless negotiated with the full range of obligations included from the beginning. Among these are the Basel Convention on the Transboundary Movements of Hazardous Wastes and Their Disposal (1989) and the Rotterdam Convention on the Prior Informed Consent Procedure for Certain Hazardous Chemicals and Pesticides in International Trade (1998). These agreements can still be amended or have additional protocols negotiated, but they essentially contain everything they need to obligate states to take action to protect the resource in question from the beginning.

A third common type of international environmental agreement is what is sometimes called a regulatory agreement.

These are agreements that create a process by which, at set intervals, decisions are made about what the actual rules will be. Fisheries agreements are almost always of this type. The agreement itself creates a fisheries commission, along with its rules of procedure, and sets a frequency of meeting. The actual decisions about what can be caught, by which states and in what ways, are made at these meetings every year or two. For fisheries agreements, this approach makes a lot of sense; information about the status of fish stocks is constantly changing, and having to go through a full renegotiation and ratification process any time fishing levels need to change would be unwieldy.

Other types of agreements use this process as well, setting up a decision-making process that decides on certain "lists" or "schedules" of what will actually be regulated. Endangered species agreements such as the Convention on International Trade in Endangered Species of Wild Fauna and Flora (CITES) (1973) and the Convention on Migratory Species (1982) make use of this approach, with the lists of species to be protected added through a regularly meeting regulatory organization. The London Convention on the Prevention of Marine Pollution by Dumping of Wastes and Other Matter (1972) follows the same procedure, with substances that are not allowed to be dumped in the ocean listed on a regularly updated list in the appendix to the agreement. (A renegotiated version of the agreement – negotiated as a protocol to the original agreement in 1996 – reverses this process, with the list containing only substances it is permissible to dump in the ocean.)

These regulatory agreements prioritize the ability to change obligations quickly, without the need to renegotiate agreements and go through a process of ratification. For that reason, many of the agreements of this type include non-unanimous voting within the commissions or organizations

Conserve , Substitute, Innovate

that make the decisions about the lists or schedules containing the substantive obligations. This allows for decisions to be taken even when some states are not in favor and thereby avoids the "least-common-denominator" outcome that can be experienced in international agreements. But states would be reluctant to join agreements that can later bind them to obligations on which they are outvoted. Sovereignty means that states cannot be bound without their consent, and, although they could consent to bind themselves to future obligations they know little about in advance, few would actually choose to do so. *"TRUMP?"*

For that reason, these regulatory agreements with non-unanimous voting tend to include provisions for states to "opt out" of rules that are created that they do not wish to follow. The agreements themselves create the process that must be followed for opting out of a rule, which usually requires doing so within a certain time period after the original decision. If one or more states opt out, an additional period of time also allows other states to decide whether to opt out, because they may no longer wish to be bound if other states are free-riding on the agreement. In some cases, if enough states refuse to be bound, the regulation ceases to bind any states.

While the provisions that allow states to opt out of regulatory decisions are sometimes controversial, they are politically necessary. Non-unanimous decision-making does allow for regulation to happen quickly and without being held up by reluctant states. Although in some agreements (for example, the early years of the International Convention for the Regulation of Whaling (1946)) it is sometimes the most important states that opt out of the obligations, in other cases few states (if any) do. These processes also allow for countries to withdraw objections (without the ability to reobject), and sometimes states can be persuaded, once the regulation is operational, that it is more valuable or less onerous than they anticipated.

The separate negotiation of new environmental agreements leads to a proliferation of agreements and institutional mechanisms to run them. In some cases this means that there may be multiple agreements that address different (or even similar) aspects of an environmental issue in different ways. For instance, the protection of species and ecosystems is subject to many international agreements, each of which protects a slightly different set of things in somewhat different ways. Among the primary efforts to address endangered species internationally is the Convention on International Trade in Endangered Species of Wild Fauna and Flora (CITES) (1973), which restricts trade in endangered species; the Bonn Convention on Migratory Species of Wild Animals (1983), which protects species that migrate on their own; the Ramsar Convention on Wetlands of International Importance Especially as Waterfowl Habitat (1971), which protects migratory waterfowl by protecting their habitat; and a variety of biodiversity agreements which protect the diversity of species and ecosystems, many centered around the United Nations Convention on Biological Diversity (1992).

At the same time, there are some international environmental problems that aren't addressed at all by international agreements. The protection of forests is one issue that is under-regulated internationally. There are some international agreements, such as the International Tropical Timber Agreement (of which there have been several over time) and other intergovernmental cooperative arrangements, such as the United Nations Forum on Forests. And forests are obviously implicated in the protection of other resources such as biodiversity and climate change and thus at least partly governed by those processes. But efforts to negotiate an international agreement to govern forests as such have failed across multiple attempts.[19]

There are a number of implications of the diffusion of international environmental agreements. One is what has been called "convention fatigue."[20] The negotiation of each agreement and the meetings and other governance processes that continue once they have been created require resources – both financial and personnel. Not all states can afford to participate fully in all agreements; the more there are, the more difficult participation becomes, especially for developing countries. Because all the financing for the operations of international organizations comes from contributions from member states, the more institutions are created by international environmental governance, the more stretched the resources for each becomes.

A second implication of numerous and overlapping agreements is the possibility for "forum shopping." Because there are likely to be multiple agreements that address aspects of the same problem, states that don't get the result they want in one political process can take their concerns to a different one in an effort to get a different result. That can occur by negotiating a new agreement altogether, as happened with the treaty to create an International Renewable Energy Agency despite the existence of an existing International Energy Agency that could, in principle, have taken up the issues the new organization was created to address.[21] Or it can involve bringing concerns that have not been met in one process to a different one. One example has been the efforts by some states to protect bluefin tuna or whales within the Convention on International Trade in Endangered Species of Wild Fauna and Flora (CITES) when the purpose-built fisheries and whaling institutions were not willing to impose strong protections on these species.[22] Forum shopping isn't inherently bad or good (it can be a useful political tactic for those seeking environmental protection), but its prevalence is an effect of the way the international governance of environmental issues works.

A third difficulty that can arise from this institutional overlap is actual conflict between the obligations or other rules of different agreements, especially when states are members of different sets of international agreements. Because international law is entered into voluntarily, there is no one set of rules that applies to all states; instead, they are obligated to the specific set of rules they have taken on. One context where this issue frequently comes up is the relationship between rules on free trade negotiated within institutions that focus on international trade and rules in environmental institutions that call for, or allow, restricting trade for environmental purposes.[23] It can be difficult enough to judge which rules apply even when all states are members of all of the relevant agreements, but it becomes even harder when some states have taken on either the trade rules or the environment rules.

Even if a number of states are all members of the same agreements, these agreements themselves, negotiated separately and for independent reasons, may conflict or fit uneasily with each other. Within the wildly successful agreements to protect the ozone layer, there were times when developing states were able both (because of the original agreement) to increase their use of ozone-depleting substances and (because of the 1990 London Amendment, negotiated in part to increase the participation of developing states) to get funding to phase out these substances – giving them the incentive to increase the use of substances for which they would then get compensation for phasing out.

Ultimately, international law to address environmental issues is a complicated patchwork of overlapping and incomplete rules, *because* states don't agree on how or whether to address different international environmental problems. The nature of the international system makes the kinds of conflicts that appear in domestic environmental politics more transparent: states that do not want to participate in protecting a particular environmental resource have the ability to avoid being bound, and efforts to bring them into agreements transform, and usually weaken, what can be done internationally. On the other hand, there may be some benefits to such a system where multiple (occasionally incompatible) rules can exist simultaneously: they allow groups of states that do agree on an issue to regulate collectively as they see fit within their areas of agreement. The fragmentation of international environmental law reflects the fragmentation of global preferences.

Non-State International Environmental Politics

The difficult politics of international environmental issues can leave spaces for other types of actors to make a difference. Different types of non-state actors can influence international environmental politics in a number of ways. Even within the process of interstate cooperation, non-state actors are influential in forming the positions of states. Non-state actors are limited in their ability to compel action, but they can nevertheless pursue options that make it more likely that states, or sub-state actors across international jurisdictions, will take steps to protect the environment. These international strategies are discussed further in chapter 4. This chapter focuses on those actions undertaken by non-state actors that are the most akin to governmental actions.

Their influence is seen most dramatically in places where states and intergovernmental organizations are failing to take action. Non-state actors can create mechanisms that operate as governing processes (albeit outside the context of governments). These approaches may eventually make state-level governing strategies more likely as well by increasing the extent to which important actors are already taking environmental action, and thus decreasing opposition to state – or international – policy.

Standards and Certification

Non-state actors can create processes that target consumers and thereby also influence the incentives of those harvesting resources or engaging in other environmentally relevant behavior. This process is sometimes referred to as "private authority."[24] It often involves what comes to be known as certification or a variety of forms of eco-labeling.

A broader context for this process is the existence of international (private) standard-setting authorities – what has been called "voluntary consensus standards."[25] The industrial world requires interchangeability of parts so that things such as screws made in one place fit in holes made elsewhere (or trains running on tracks in one country are able to continue on tracks when they get to the next country). Those standards are set not by intergovernmental organizations but instead by gatherings of nationally based (and usually private) standards-setting organizations. One of the important standard-setting organizations relating to environmental governance is an independent NGO, the International Organization for Standardization (ISO), with a membership of 163 national standards bodies. It produces standards for industry and commerce generally but has also created a set of voluntary standards (ISO14000 and 140001) that provide guidance for environmental management and a process by which organizations can be certified as working within these standards.

In international environmental politics, this broader approach to environmental certification began in the forest sector – as efforts to produce an international forestry agreement have largely failed over time and global deforestation is rampant. Initially, forest-related NGOs created their own auditing processes in their attempts to track the environmental impacts of deforestation. This happened in the context of more formal intergovernmental action that measures and organizes deforestation but doesn't mandate forest conservation. The UN Food and Agriculture Organization monitors the increasingly dramatic loss of tropical forests. The International Tropical Timber Agreement (ITTA) is a commodity agreement that (unlike more traditional international environmental agreements, which are intended to operate indefinitely) is periodically renegotiated to last for a set length of time; it creates goals for sustainable harvests but does not mandate them.

When the governing body for the ITTA refused to consider a labeling initiative to track forest products, NGOs decided to take matters into their own hands. First, the Rainforest Alliance created a program (called "SmartWood"), and other organizations were created by woodworkers who wanted to be able to ensure that the woods they used were harvested sustainably. The fact that forest products are largely traded internationally opened both the need and the opportunity to try to track timber products as they traveled and to ensure, for those who were interested, access to sustainably harvested wood products. Businesses, discussed further in chapter 4, wanted to offer such options to their environmentally interested consumers but needed to be able to track and guarantee these products. These initial private efforts at certification eventually gave way in 1993 to the creation of the Forest Stewardship Council (FSC).[26]

The FSC is an NGO made up of environmentalists, forest industry professionals, and others, which has a complex and transparent decision-making process. It sets the standards a forestry operation has to meet in order to be FSC-certified as sustainable. Those who own or harvest in a forest need to apply to become certified; at that point an FSC-accredited certification authority checks to ensure that the operations meet the designated criteria and, if so, allows its products to bear the FSC label. Among those criteria is supply-chain monitoring, which allows the certified company to ensure that its products can be reliably tracked as they are moved internationally. Nearly 200,000 hectares of forest across eighty-four countries are certified under FSC's system.[27]

This approach has been replicated in the context of oceans with the creation of the Marine Stewardship Council to certify fish that are sustainably harvested. This organization began in 1997 as a collaboration between an NGO and a major seafood industry conglomerate but has since become an independent organization. As with the FSC, it sets standards that fishing operations must meet in order to be certified, and certified seafood is designated as such. Similar organizations and processes have sprung up around coffee and other consumables. In some cases other organizations have tried to offer competing certification or labeling. In the case of forests, an industry-led organization – the Sustainable Forestry Initiative – operates in this manner, with predictably more industry-friendly sustainability standards.

Across issues, one major target of certification processes is consumers, who allow such organizations to make decisions on their lumber or seafood based on its sustainability. And, by doing so, certification gives incentives to the suppliers to choose to produce in a sustainable manner and go through the effort to be certified as such, because it allows them to reach this set of consumers who might be willing to pay a price premium for environmentally benign products.

These certification processes have the greatest effect where they go beyond simply providing information to consumers and become integrated in purchasing or management decisions by other entities. Stores may agree to stock only MSC-certified wild seafood (as Loblaws in Canada vowed to do by 2013 and Walmart in the United States by 2011). MSC also grants chain-of-custody certification to distributors, which certifies that all the seafood they sell is MSC-certified. In those contexts, consumers do not need to seek out, or even know about, certified seafood; any products they buy within a given store will be certified as sustainable. Similar processes can influence the extent of wood certification. Building standards, including those via other certification processes for green building design (such as Leadership in Energy and Environmental Design (LEED)), may require or incentivize using certified lumber. These are thus entirely non-governmental endeavors that affect the protection of resources across borders without ever making binding national or international laws to do so.

Other Standards

There are other ways in which NGOs can have an effect on international politics of the environment. Many of these are illustrated in the context of climate change, where international policy has been difficult, slow, and contentious, especially with the reluctance of the United States to participate in major international climate change processes.

Ironically, one way to work around state reluctance in international policy is to operate below the state. The organization initially called the International Council for Local Environmental Initiatives, which now simply goes by the acronym ICLEI, has been instrumental in persuading municipalities and other sub-state entities to undertake pledges to reduce their contribution to climate change and assisting them in mapping out ways to do so. The organization's Cities for Climate Protection (CCP) program guided cities through the measurement, commitment, planning, implementing and monitoring for pledges to reduce greenhouse gas emissions. Currently more than 3,500 municipalities and other sub-state entities participate in ICLEI-led environmental initiatives.[28] There are numerous other NGOs that manage efforts to get local entities to take actions affecting climate change, a truly global issue.

Non-state actors have also been central players in other international efforts to manage climate change. These include efforts to keep track of and measure greenhouse gas emissions commitments. ICLEI has a voluntary "reporting platform" for local governments to publicly report and monitor commitments they have made to reduce their contribution to climate change.

A more central decision-making role has been played by two more traditional environmental NGOs in creating the *de facto* standard for how greenhouse gas emissions are accounted for by firms. Any reduction in greenhouse gas emissions needs to be calculated – it's not something that is generally directly measured and, in any case, if there is either an obligation or a commitment to reduce, it's important that everyone doing so has the same method of calculating the reductions. This is true of formal government commitments to reduce emissions but also true of businesses or other entities that pledge or are otherwise required to reduce their emissions. In this case, the process that has become the global standard was created in a collaboration between two NGOs, the World Resources Institute (WRI) and the World Business Council for Sustainable Development (WBCSD).[29]

NGOs have played other formal roles in intergovernmental climate institutions. An important one is via the Clean Development Mechanism (CDM), a funding mechanism within the United Nations Framework Convention on Climate

Change process. Under this process, developing countries can receive funding to meet the "incremental costs" of actions they take to reduce greenhouse gas emissions. That process, therefore, requires an independent entity that can calculate and attest to what that additional cost is. The CDM maintains a list of "designated operational entities" that are able to produce those independent evalutions; many of these are NGOs.[30]

NGOs play quasi-governmental formal roles within existing international governance processes. That is true of NGOs such as TRAFFIC (originally Trade Records Analysis of Flora and Fauna in Commerce), a conservation organization founded in 1979 to track wildlife exports and other threats to endangered species. Its reports have been influential enough to be adopted within decision-making bodies of CITES.[31] An even more compelling example is the International POPs Elimination Project (IPEP), which was created to assist countries implementing the Stockholm Convention on Persistent Organic Pollutants. Intergovernmental organizations such as the United Nations Environment Programme (UNEP) and the United Nations Industrial Development Organization (UNIDO) have provided funding to this otherwise non-governmental organization, assisting states with intergovernmental rules.[32]

Conclusion

The international political process requires a kind of persuasion and compromise that often leads to weaker rules than environmentalists and scientists would advocate, because international rules apply only to those states that decide to take them on, and most states need to participate for international rules to be effective. Despite those difficulties, international rules have been negotiated to address a variety of international environmental problems, and some have had notably beneficial effects.

Also important are intersections between domestic and international environmental politics. States that have acted on an issue domestically may be more inclined to lead on that issue internationally, or at least not resist taking action. That international willingness can come because those states have already concluded that the problem is worth addressing, or that they have discovered that taking action to address it has been less onerous than they initially expected. It may even be that regulating at the state level for an environmental issue that can only be addressed fully if other states in the world also take action prods the state that initially took action to push the issue onto the international agenda. That is particularly likely if the initial regulation creates a cost for internationally competing industries within that state and the industries in other locations do not have to bear that cost. The regulated industries will want their competitors to bear the same cost in order to even the economic playing field, and they will therefore likely lobby their own governments to work for international action.[33]

At the same time, international action on an issue may open up space for domestic action by those states that have not already engaged with it politically. Negotiating internationally about an environmental problem engages scientific research that can illuminate its dangers and the options for addressing it. International agreements, especially recent ones, are likely to come with technical and sometimes even financial assistance for developing countries, which can make taking domestic action possible. And, of course, once states agree to international environmental rules, they then are obligated to take whatever domestic action is required to make the necessary changes to meet those obligations. International environmental politics, though operating differently than domestic environmental politics, thereby engages with it in important ways.

An issue such as climate change demonstrates both the difficulty and the creativity of international environmental politics. For all the reasons discussed elsewhere in this book, climate change is as difficult an environmental problem to address as can be imagined: it is a global collective action problem in which activities central to industrialization and agriculture in the present cause long-term environmental harm over long periods of time in places unrelated to where emissions originated. While the basic scientific processes are understood, there is nevertheless much uncertainty about details, in part because of the long time-horizons involved. Many of the most powerful states have been reluctant to commit to emissions reductions significant enough to make a difference in the environmental problem. When the initial international efforts (through the United Nations Framework Convention on Climate Change and the Kyoto Protocol) failed to yield emissions reduction commitments from the United States and the major developing states, international negotiators tried again, with the Paris Agreement, designed to bring these states in, with some success. In and around these intergovernmental efforts, NGOs and sub-state entities have worked on their own commitments to augment or substitute for intergovernmental action. Though we are far from the changes needed to prevent major climate catastrophe, the political efforts continue.

CHAPTER 6

Engaging with Environmental Politics

It is easy, at this point, to be disheartened about environmental politics. There are so many ways politics has failed to protect the environment, and it may feel as if the deck is stacked against political processes in the effort to solve environmental problems. Several concluding observations are thus important. First, although we may decry the inability of political structures – as of yet – to respond sufficiently to global climate change, the major environmental issue of this generation, we should also marvel at the extent to which political processes have made great strides in addressing environmental problems worldwide. Second, we should consider whether it is the political processes that deserve the blame for lack of progress on some environmental issues, or whether politics is simply the representation of an underlying lack of consensus about the best ways to make social decisions about environmental problems. Finally, we should consider the alternatives: if we do not attempt to address environmental problems through political processes, what other options are there? To what extent will they fall prey to similar – or worse – difficulties?

In short, environmental politics may be a bit like the phrase about democracy attributed to Winston Churchill: it is "the worst form of Government except all those other forms."[1] Environmental politics may be an inefficient and imperfect tool for efforts to protect the environment, but it has advantages over other options. And since political decisions with environmental effects are taking place all around us, ignoring the politics of the environment will likely result in worse outcomes than engaging with political processes will.

Environmental Progress

It is easy to see the ways that the political process is failing the environment. A recent United Nations assessment of the global environment finds dramatic environmental damage. Between 6 and 7 million premature deaths and economic losses of $5 trillion annually are caused by air pollution. We are in the middle of a major loss of species and biodiversity so dramatic that some compare it to the type of extinction event that wiped out the dinosaurs. Estimates suggest that vertebrate species abundance has declined by 60 percent in the last forty-five years; in addition, 42 percent of terrestrial invertebrates, 34 percent of freshwater invertebrates, and 25 percent of marine invertebrates are in danger of extinction. Plastic pollution of the oceans is found throughout the entire system of world oceans, at all depths. The quality of fresh water worldwide has decreased significantly in the last twenty-five years, primarily because of pollution by chemicals and heavy metals.[2] Fisheries depletion across the world's oceans is severe, with more than 33 percent of fish stocks globally unsustainable and another 60 percent of stocks incapable of sustaining any increase in fishing effort.[3] Global populations of large predatory fish (those species such as tuna and swordfish that are subject to international management) are at a mere 10 percent of their pre-industrial exploitation levels.[4]

Global climate change is the most dramatic unsolved global environmental problem, and, as our understanding of the problem improves, earlier predictions have turned out to have underestimated its severity.[5] The most recent report by the generally cautious Intergovernmental Panel on Climate Change, the international body of scientists studying all aspects of climate change, anticipates that current climate trends will lead within the next two decades to the destruction of all coral reefs, to global wildfires and heat waves, and to dramatic increases in food insecurity, among many other effects – not to mention the loss of hundreds of millions of human lives.[6]

But it is important not to lose sight of the dramatic environmental improvements, from the most local to the most global, that have been brought about, in large part, by political processes. Indoor air pollution, one of the major contributors to poor human health, has decreased by more than a third, and annual premature deaths attributed to indoor air pollution globally have fallen from 3.7 million in 1990 to 2.6 million in 2016,[7] despite the increase in population in many of the countries most at risk for this type of pollution.

Other types of air pollution have also dramatically decreased in parts of the world. In the United States, as in many industrialized countries, emissions of the five types of air pollutants that directly affect human health have decreased by nearly two-thirds since 1970, despite a growing population using greater transportation and more energy.[8]

Despite increasing water pollution, the percentage of people with access to clean drinking water has increased dramatically, with 91 percent of the world's population having what the United Nations refers to as "improved drinking water sources." In addition, 2.1 billion people have gained access to sanitation since 1990.[9]

And although species decline is major and pervasive, some important species – elephants, condors, and manatees (among many others) – have been successfully brought back from dramatic depletion. The rate of bird extinction has decreased.[10] Countries have dramatically increased the number and size of protected areas both on land and in the oceans. Marine protected areas now constitute at least 7 percent of the ocean (representing an area larger than North America).[11] Terrestrial protected areas in Latin America and the Caribbean now constitute nearly a quarter of the entire land area of that region.[12]

At the international level, there are many success stories. One of the most impressive involves the set of international agreements to protect the ozone layer, discussed in chapter 5. Ozone depletion shares with climate change many of the characteristics that make it difficult to address: its effects are diffuse and felt over long periods of time, and states pretty much everywhere in the world needed to change important aspects of industrial activity to make a difference. A set of carefully negotiated international agreements allowed international political processes to take account of new information and assist developing states and ultimately resulted in a complete worldwide phase-out of most uses of ozone-depleting substances. The outcome is a recovering ozone layer and a transformation of industrial activity at a cost far lower than most anticipated.

International cooperation has produced many other successes. Acid rain in Europe has dramatically declined, with emissions of most relevant pollutants decreasing by between 40 and 80 percent and soils and freshwater recovering from previous acidification.[13] Oil tankers worldwide have been redesigned in ways that make oil pollution of the oceans much less likely; despite dramatically increased transportation of oil, the number and volume of oil spills from tankers has been reduced from a high of nearly 650,000 metric tons annually in the late 1970s to 7,000 metric tons annually in 2017.[14] Other forms of operational oil pollution have declined as well.

These advances happened in large part because of political processes. Although environmental protection becomes more likely as societies become wealthier, that process is not automatic. It requires a set of social decisions and political action. Access to clean water and sanitation necessitates a level of infrastructure that can be provided on a large scale only by governmental action. A decrease in air pollution comes not from the altruism of industry but from rules that require reductions in emissions, along with incentives that can assist with the development or adoption of new technologies. Poachers rarely decide on their own to stop harvesting threatened species; instead, rules, monitoring, and enforcement increase the cost of continuing their species-threatening behavior. Even when businesses or industries decide on their own to make environmental improvements, the spectre of potential regulation if their behavior doesn't change often contributes to those decisions. The role of politics is even clearer at the international level, since treaties are negotiated and then implemented by governments. Most environmental improvement comes, at least in part, from environmental politics.

Politics as the Reflection of Society

Environmental politics is indeed a contentious process. But if we return to the definitions in chapter 1, we'll remind ourselves that politics is the process of societies making decisions about policy; it is about how to allocate resources of all types across society. People studying environmental politics may begin with the belief that it is important to protect natural resources and prevent pollution, whether for the wellbeing of people or of species or ecosystems more broadly. But these are not the only values, even for those who hold them. The politics of the environment involves legitimate tradeoffs between (some kinds of) jobs and environmental protection, between convenience and health, and between short-term and long-term benefits, among many others. Pretending that coal miners won't actually be put out of work or oil companies won't lose money if society gets serious about clean air or preventing climate change is naïve and misses the sources of political opposition to efforts to address environmental problems.

There are some aspects that can make these contentious elements more difficult for making social decisions about the environment than about other issues. The long time-horizons of many environmental problems fit uneasily with political decision-making structures in which politicians are elected or kept in office based on the short-term benefits they are able to create. The common pool resource nature of environmental issues means that collective action problems are made worse by subtractability, allowing free-riders to undermine the ability of those who do participate to solve a problem. The distance between where problems originate and where they are experienced means that winners and losers from environmental action may be in different political jurisdictions. In other words, fixing environmental problems, through political processes or in any other way, is legitimately difficult.

Ultimately, then, environmental politics is contentious because there are no easy answers. There are some elements of political processes or specific environmental issues that can make any one effort at environmental politics more or less difficult. But protecting the environment is always going to require political tradeoffs and persuasion.

Alternatives to Environmental Politics?

If political structures are misaligned with environmental structures in a way that makes it difficult to address environmental problems through politics, what are the alternatives?

Some people argue that what is required is a change in our collective consciousness. Environmental issues are a moral imperative, and perhaps the way to address them is to persuade individuals to care enough about the environment to change their behavior collectively towards actions that have a less problematic effect on the environment.

There are some good arguments for making the environment a moral issue. In polls across the world most people indicate that they are willing to "do what is right for the environment," even at a personal cost.[15] It *is* important to value the wellbeing of human populations harmed by actions taken in search of economic profit. In many cases, the most vulnerable populations are the ones least responsible for creating environmental problems and the ones who suffer the most from them. In addition, the lives of non-human species, or even ecosystems, may be worthy of moral consideration; they are affected by human actions and can also suffer, at least in some cases.

People who act out of moral concern may be more willing to make changes to protect the environment. Moreover, if people make profound rather than simply convenient individual changes, those changes are likely to persist and to expand outward to influence related areas of life.[16] People who become vegetarian no longer need to engage in an individual calculation each time they face the option of eating meat; it's simply not an option. (The same can be true of forming new individual habits; if you regularly bring your reusable bags to the grocery store or walk to work, you will likely continue to do so.) If you take an action because you genuinely believe that it is the right thing to do, you are more likely to be able to keep doing it even when it becomes inconvenient.

There are some downsides, however. People don't like being told what to do, and especially being told that they are behaving in morally incorrect ways; they may react by persuading themselves that their existing behavior is morally acceptable and thereby be less likely to change. People tend to engage in a collective moral balancing (sometimes referred to as "moral licensing") in which taking an inconvenient action for ethical reasons might make them willing to allow themselves to make the next decision in a less morally concerned way.[17] So making protection of the environment a moral issue can backfire.

An alternative strategy for focusing on individuals is to reach them through fear. Environmental problems such as climate change are genuinely scary, and the more we learn about their implications the more fear-inducing they may be. Although fear, which is a hallmark of many public health campaigns, might persuade some people to change their behavior, it can also backfire, especially for threats such as climate change that people cannot individually control. This effect has been demonstrated repeatedly in psychology experiments. Frightening details may lead people to ignore information or to emphasize the extent to which a problem is unlikely to apply to them. Studies of "dire warnings" presented about climate change suggest that they make people not only less likely to believe such information than others who did not receive fear-inducing particulars, but also less likely to believe than they did before receiving these messages.[18]

Fear and moral suasion focus on achieving changes in individual behavior within existing political parameters. Perhaps what we need is radical change rather than business as usual – a revolution – anarchy, socialism, or some other approach that dramatically upends how we govern ourselves. Again, there's an argument to be made for revolutionary change rather than the slow slog of existing political processes. But how will that revolutionary change come about? Some activists argued in the United States that the election of Donald Trump would cause conditions to worsen sufficiently to bring about a revolution. But even if this process has moved opposition politics leftward (and the answer to that question is unresolved), responses are still taking place within the existing political system that is now slanted further away from environmental action. And genuine country-changing political revolution is extremely rare, especially within advanced industrialized democracies, and extremely disruptive of people's lives whenever it does occur.

A Defense of Environmental Politics

There are nevertheless enormous advantages to working through political processes to address environmental problems. Most of the environmental issues people are most concerned about are widespread and pervasive. Many people would have to change many types of behaviors in order to make progress in addressing them. Politics may be slow, but it is faster than the alternatives. (Imagine having to persuade 7.5 billion people individually to make different decisions about environmental behavior.) Politics may be parochial – focused on the needs of the particular political community involved rather than the world as a whole – but it is able to create widespread change in that community.

Moreover, political – social – decisions have the ability to change social and economic structures in a way that individual decisions do not. No amount of individual recycling activity will reduce the amount of wasteful packaging in the way that mandated waste fees might, or indeed requirements that producers take back their packaging. Reducing your electricity use (or even creating a house off the electric grid) will not magically cause renewable energy to become available for everyone, but laws requiring utilities to produce a certain amount of renewable energy could.

Similarly, a focus on individual action misses the social privilege that may make more environmentally friendly options available primarily to those with greater wealth or other social advantages. In the absence of requirements, subsidies, or other incentives, making a better environmental choice will frequently be more costly or more difficult than making a worse environmental choice. That situation can be traced back to the fact that environmental problems come from externalities. Because the externalities don't themselves come with a cost, acting in a way that doesn't create externalities probably does. So if it is up to individuals to change their behavior, it will be the easiest to do for those who have the most privileged places in society.

If education, moral suasion, or even fear have beneficial effects on environmental behavior – and the jury does appear to be out on those questions – they can also serve as tools in the process of environmental politics. People who are morally committed to environmentalism are the ones likely to put in the effort to coordinate the collective action necessary to persuade people – most of whom genuinely do want better environmental conditions – to take the required action to achieve policies that will impel better behavior from everyone. Educating people about the urgency of environmental action and the benefits that come from addressing environmental problems may make them more receptive to new rules created within a political process. Individual action alone may be insufficient, but when a proportion of the people in a community have already changed their behavior to benefit the environment, the political process will encounter fewer roadblocks to mandating change.

A final argument for engaging in politics to help address environmental problems is that political processes are always going to be ongoing, and they are always going to have environmental implications. People who are concerned about the environment have no choice but to engage in political processes in defense of the resources or future generations that are not able to participate in political processes themselves. If those of us who are concerned about the environment absent ourselves from politics, there will be no one to argue for environmental concerns in those discussions, and short-term economic interests will prevail.

How to Protect the Environment, Politically

This book is about the processes of environmental politics rather than any particular policy choices that could be pursued. Nevertheless, the political processes introduced here suggest approaches that may be more – or less – fruitful in protecting the environment and also some things about which types of political systems or processes may be most amenable to attempts at environmental protection. There are options for productive political action on the environment and ways to make beneficial environmental outcomes more likely.

The overview of environmental politics in chapter 1 suggests that those who are concerned about the environment should take seriously some of the characteristics of environmental problems that make them difficult to resolve. We can take comfort in knowing that, because environmental problems are externalities, no one intends to create them. Find ways for people, communities, businesses, or other actors to accomplish their underlying goals in ways that don't harm the environment. And acknowledge that those underlying goals matter: businesses seek profit; people have important needs beyond a clean environment (including jobs and education and the ability to provide for their families). If there is a potential backlash to environmental policy, it may be managed or avoided by anticipating concerns and then designing policy around that. Will gas taxes fall primarily on poorer people less able to rely on alternate commuting options? A tax-and-dividend program (where tax receipts are used for refunds based on income level) can help. Consumers may be less opposed to policies that reward them for good behavior than those that punish them for problematic behavior. Framing and policy design is especially important because addressing environmental problems often requires some kind of loss or sacrifice initially; figuring out how to make any changes less difficult for underlying goals people face can be key.

Environmental politics requires collective action, which can be difficult for environmental problems where the benefits of change are diffuse and may be felt far in the future, perhaps even by people other than those taking action. At the same time, the primary costs of such steps are often concentrated on larger actors – business or industry – that find it easier to organize against them. When environmental benefits accrue to everyone, regardless of whether they participated in making them happen, it is easy to prioritize other concerns and free-ride on the environmental activism of others. (And when environmental outcomes seem unlikely, it can be hard to persuade people to spend their scarce resources pushing for them.)

Find ways to make collective action easier. Make it more convenient for people to come to meetings (provide childcare or interpreters, or schedule them at times and locations that make it easier to participate). Provide selective benefits – whether they be tangible things such as food at meetings, access to specific resources (a T-shirt or access to hiking trails), or simply making your organizing events fun and inspiring – that give people a reason to participate in addition to caring about the policy goal.

Chapter 2 suggests that science will neither save nor doom us in engaging in environmental politics. On the one hand, environmental problems are often created in part because we don't know that our actions will cause harm. Stopping to ask the question of what potential environmental harm could come from our activities before creating a product or process would be an excellent general approach. Improve scientific research – support it, amplify its findings – to increase general awareness of environmental problems and provide full information that can inform political processes. Work to increase scientific literacy and awareness of biases in approaching information and evaluating risk. Ensure that research provides the information most useful to political processes.

Most important, though, is to recognize that, while it is often useful to have as much information as possible about the parameters of an environmental problem, and the costs and benefits of the actions that create or mitigate it, having this information is inevitably insufficient. No matter how well communicated, science cannot make the social decisions that environmental politics requires. If you want to argue for environmental protection within the political process, then, the task is to make the case that it is worth the tradeoffs. Argue in favor of the values of health and wellbeing for the people who will be protected – even at some economic cost to others – by preventing or fixing environmental problems. Make the case that ecosystems or species are worthy of protection (perhaps even at a cost to people). Make it clear that short-term concentrated costs may yield medium-term collective benefits that outweigh those costs (to society, if not to the individual industries that bear the cost). And engage in the politics that help give voice to the people who benefit from environmental protection.

Examining how different country-level political structures (chapter 3) and actors (chapter 4) contribute to environmental politics is important, but it can be hard to draw immediate recommendations for political action: after all, even if democracy turns out to be better than authoritarianism for political action to protect the environment (for example), vast changes in political structures are unlikely within a short time-frame, and people do not frequently get to choose the locations where their political activism takes place.

Democracy is a useful example, though. Even if there may be some instances in which authoritarian governments that happen to have an environmental aim can work quickly to enact policy to support it, the one clear advantage of democracy for the environment is in the political process. People have the ability to influence political change more directly, and democratic systems are likely to be more responsive to the interests of their populations. And democracy is valuable for reasons other than environmental action; protecting rights and responding to the concerns of communities is valuable no matter what those rights or concerns are.

So one important approach is to work to increase how democratic political jurisdictions are, no matter their starting level. If anything, the very recent trend is in the wrong direction. After many years of increasing democratization, in 2016 more countries decreased their level of democracy, as measured by the Economist Intelligence Unit, than increased it.[19] Protecting democracy is essential. Work to overcome corruption, which makes environmental protection less likely and is problematic in its own right. Improve economic and political equality, which, again, has its own benefits but also increases the likelihood of positive environmental outcomes. The downsides of democracy include a potential outsize influence of concentrated actors – corporations or industry – which are the primary polluters and users of resources. Increase equality, which is beneficial on its own and will also improve the influence of those most likely to be harmed by environmental damage relative to those most likely to cause it. Of particular note is the need to improve the situation for the most vulnerable populations: those discriminated against by race, religion, or language or other background and who are likely to be the main experiencers of environmental injustice. Working to improve the situation of those populations is good for society generally and will also give voice to people who will advocate in favor of environmental protection.

Within democracies, the details of what type of political system exists suggest different strategies for approaching environmental politics. Within presidential systems such as that of the United States, with separation of powers and majoritarian electoral rules, third-party candidacies are generally doomed. Supporting green parties in this context is as likely to be counterproductive – splitting the vote among further left traditional (e.g. Democratic) party candidates and increasing the electoral chances of those opposed to environmental action. That suggests a strategy of working to move the more environmental of the existing major parties towards greater support for the environment rather than pushing candidates outside the traditional parties. (The only exception is likely to be the local level, where serious organizing may make alternative candidates viable, but that is still likely to be the exception rather than the rule.)

The alternative is to work to change electoral rules; in the United States, some localities – and even states[20] – are adopting ranked-choice voting (also known as "instant run-off voting"). This allows the reallocation of votes for candidates who receive only a small number of votes and thus makes voting for those outside of a traditional party structure more viable even in majority-requiring contexts. It ultimately allows for increased electoral opportunities for these contenders, since voting for them does not entail "throwing away" the ability to help elect the winning candidate.

Separation-of-powers frameworks also create opportunities to use the legal system to force environmental action when the party that implements laws is different from the party that makes them. So work to ensure that the judicial branch has professional and environmentally concerned jurists when an environmentally sympathetic party holds power over judicial appointments, because these appointments generally outlast the term of office of those who appoint. The same is true for bureaucracies, which should be staffed with competent and sympathetic supporters of environmental action whenever possible.

Green parties are more viable in parliamentary systems, especially those with larger numbers of parties or primary parties that may not have majority electoral support. Similarly, for proportional representation systems, it is key to get the names of environmentally concerned candidates high up on the lists and to work broadly for votes for that party, whereas for single-member district systems throwing support behind a specific environmental candidate may be a good strategy. The broader lesson for environmental politics is to pay attention to the electoral rules – how candidates are elected – to determine which candidates to run or to support in an election.

Federalism, likewise, provides opportunities. If national politics is not supportive of environmental outcomes, work at the level of the state or province to craft environmental policy that can at least cover some portions of the country. If that policy succeeds, it may be more appealing to other political units or even nationally, and sub-national jurisdictions that have already taken environmental action are more likely to push for, or at least not resist, national political action. Use the courts, which in many places can hold individuals – or even governments – to the laws the political process has created.

It is also possible to influence environmental politics through other actors. Join or contribute to environmental interest groups. Push business and industry to choose environmental action (and choose to support businesses that do). Recognize that finding common ground between businesses and environmentalists can create effective political action, and push for environmental rules that might in some other ways benefit the industries that would be regulated, decreasing the political opposition to regulation.

Various forms of media present both opportunities and dangers. Work to increase the scientific literacy of both the general public and the news media and call attention to false balance in how environmental information is presented. Use social media to get out environmental messages and to expose misinformation, misleading presentations (including cherry-picked data), and outright fake news. The increasing extent of media bubbles, in which people get information from only like-minded sources, presents a real danger to how environmental issues are perceived, but it can be extremely difficult to interrupt this information isolation.

Figuring out how to access international politics, as discussed in chapter 5, can also be a challenge, but understanding how politics works at this level is central both to figuring out what may be politically possible and to pushing states and other actors towards more effective efforts to address international environmental problems.

Countries are still the most important actors at the international level, so an important way to influence politics internationally is to work to persuade states to support and pursue strong international action. States are more likely to take action internationally when they have committed to addressing an environmental issue domestically,[21] so work at this level on internationally relevant environmental issues can have an international effect. Bureaucracies matter internationally too: the agencies or organizations that work to implement political agreements made by states can play an important role in the success of environmental action. Work for these organizations, contribute to them, or provide the scientific or social scientific research on which their work depends.

There are downsides to the multiple and overlapping rules that get created through the complex process of international environmental politics, but the upside is that they create multiple pathways to take action. If one international agreement or process does not protect a resource sufficiently, try a related agreement that might offer a different avenue for protection. Or work for certification processes or sub-state commitments to act on internationally relevant issues on which international action isn't forthcoming.

Remaining Questions

There are many unanswered questions about the politics of the environment and excellent topics for research that could test hypotheses or provide evidence to support or overturn our understandings of environmental politics. Efforts to understand the role that different political structures play in environmental outcomes are, surprisingly, in their early stages. It's remarkable that we still don't have consensus on the role that major factors such as democracy or even wealth play on environmental outcomes.

One of the big advantages of studying environmental politics is that variation across different political units, levels, or issues creates opportunities for studying what factors matter most. If you want to understand more about what elements of state structure or which state characteristics influence how or how well states implement environmental policy, examine how different countries tackle the same environmental issue. A study could examine how countries with different levels (or different types) of democracy address climate change, or how successfully countries with different levels of inequality tackle water pollution or any type of environmental issue. States or provinces within federal systems provide a similar laboratory: when different sub-national units have the rights to regulate in different ways, what influences whether, or how, they decide to act?

At the international level there are many issues that are addressed through multiple versions of one type of regulatory approach across different regions. There are more than twenty regional fishery management organizations (RFMOs), the governing bodies that determine fishing rules for different fish species or regions. There are more than eighteen regional seas programs governing pollution and resource use within different semi-enclosed seas; they vary in region, institutional arrangements, and obligations of member states. The number and size of marine protected areas is changing dramatically, and they vary in what activities they allow to take place within protected areas.

The same type of approach can help illuminate how different levels of political structures address environmental issues. What is different, or more or less successful, about how a city, country, or region tackles an issue? What determines the level at which the issue will be addressed politically?

Conclusion

Addressing environmental problems through the political process is a difficult thing to do. Environmental problems have characteristics that make them hard to address in any context: they are unintended consequences of other things that people value; while it doesn't cost anything to create problems, it probably does, at least initially, to prevent them. Fixing or preventing environmental problems, therefore, comes with a social cost: it makes things harder, or more expensive, for people who may, therefore, oppose taking that action. The costs of change are felt in the present and the benefits generally in the future, perhaps by people quite far away.

The political process compounds the difficulties these characteristics present. Politicians represent their district or country, and environmental problems often cross political boundaries. Undertaking costs here in order to prevent damage elsewhere is a difficult political choice. Elected officials also have reasonably short time-horizons, concerned with providing benefits (or avoiding costs) that accrue before the next election that could turn them out of office. They are likely responsive to constituents that have the greatest level of political power.

And yet, political processes have a major influence – for good or for bad – on the condition of the environment. Politics is the way communities make decisions about social values. How important is economic growth, or provision of energy, or species conservation, or pollution prevention? Who will suffer what costs or what environmental damage? There is no alternative to engaging in this process of social choice; science will not tell us the right thing to do environmentally (or economically); it can only help us better understand the tradeoffs. Deciding among those tradeoffs is the job of politics.

Understanding the systems and structures through which these political decisions are made is key to understanding how and when environmental problems are prevented or addressed. Aspects of the environmental issues themselves matter. How widespread is the activity that contributes? How well do we understand its causes and effects? Where and by whom are those effects felt? How difficult or costly would it be to change?

Aspects of the political process also matter. How democratic is the country in which the problem takes place? Within democracies, how do the electoral rules influence who the policy makers are? How unequal is the society in terms of access to the political process? How politically active and well organized are the people who create, or suffer from, the problem?

Or is it a problem that can be addressed successfully only at the international level? Which countries are the winners and losers environmentally or economically from the problem? How do the domestic political situations of the most important states – particularly in contributing to the problem – influence their willingness to address it collectively?

Understanding environmental politics can help us understand both why it is difficult to prevent or fix environmental problems and what the most effective routes towards creating environmental change are. We need to understand the elements of environmental issues or political structures that create that difficulty so that we can work either to change those elements or to structure a political strategy to circumvent them. What happens politically will affect the environment and the wellbeing of the people and other species that depend on it. Figuring out how to engage environmental politics for the benefit of the planet and its inhabitants is therefore a worthy undertaking.

Notes

CHAPTER 1 DEFINING ENVIRONMENTAL POLITICS

1 Ronald H. Coase, "The Problem of Social Cost," *Journal of Law and Economics* 3 (1960): 1–44, at p. 13.
2 Robert Costanza, "Social Traps and Environmental Policy," *BioScience* 37/6 (1987): 407–12.
3 Julian Simon, *The Ultimate Resource* (Princeton, NJ: Princeton University Press, 1998).
4 Paul Sabin, *The Bet: Paul Ehrlich, Julian Simon, and Our Gamble over Earth's Future* (New Haven, CT: Yale University Press, 2013).
5 "Oil: High Costs, High Stakes on the North Sea," *Time*, 29 September 1975.

CHAPTER 2 UNCERTAINTY AND SCIENCE

1 Marc A. Levy, "European Acid Rain: The Power of Tote-Board Diplomacy," in Peter M. Haas, Robert O. Keohane and Marc A. Levy, eds, *Institutions for the Earth: Sources of Effective International Environmental Protection* (Cambridge, MA: MIT Press, 1994).
2 M. J. Molina and F. S. Rowland, "Stratospheric Sink for Chlorofluoromethanes: Chlorine Atom-Catalyzed Destruction of Ozone," *Nature* 249 (1974): 810–12.
3 Discussed in Matthew Paterson, *Global Warming and Global Politics* (London: Routledge, 1996), pp. 17–21.
4 Richard J. Hamilton, Glenn R. Almany, Don Stevens, Michael Bode, John Pita, Nate A. Peterson, and J. Howard Choat,

"Hyperstability Masks Declines in Bumphead Parrotfish (*Bolbometopon muricatum*) Populations," *Coral Reefs* 35/3 (2016): 751–63.

5 Michaël Aklin and Johannes Urpelainen, "Perceptions of Scientific Dissent Undermine Public Support for Environmental Policy," *Environmental Science & Policy* 38 (2014); 173–7.

6 Seong-lin Na and Hyun Song Shin, "International Environmental Agreements under Uncertainty," *Oxford Economic Papers* 50/2 (1998): 173–85.

7 Lawrence E. Susskind, *Environmental Diplomacy: Negotiating More Effective Environmental Agreements* (Oxford: Oxford University Press, 1994), pp. 66–7.

8 S. Lichtenstein et al., cited in Lola L. Lopes, "Risk Perception and the Perceived Public," in Daniel W. Bromley and Kathleen Segerson, eds, *The Social Response to Environmental Risk* (New York: Kluwer Academic, 1992), p. 60.

9 Rose McDermott, *Risk-Taking in International Politics: Prospect Theory in American Foreign Policy* (Ann Arbor: University of Michigan Press, 1998).

10 Daniel Kahneman and Amos Tversky, "Prospect Theory: An Analysis of Decision under Risk," *Econometrica* 47/2 (1979): 263–92.

11 Daniel Kahneman, Jack L. Knetsch, and Richard H. Thaler, "Anomalies: The Endowment Effect, Loss Aversion, and Status Quo Bias," *Journal of Economic Perspectives* 5/1 (1991): 193–206.

12 Ziva Kunda, "The Case for Motivated Reasoning," *Psychological Bulletin* 108/3 (1990): 480–98.

13 Brendan Nyhan and Jason Reifler, "When Corrections Fail: The Persistence of Political Misperceptions," *Political Behavior* 32/2 (2010): 303–30.

14 Dale Griffin and Amos Tversky, "The Weighing of Evidence and the Determinants of Confidence," *Cognitive Psychology* 24/3 (1992): 411–35.

15 James Flynn, Paul Slovic, and Chris K. Mertz, "Gender, Race, and Perception of Environmental Health Risks," *Risk Analysis* 14/6 (1994): 1101–8.

16 Aaron M. McCright and Riley E. Dunlap, "Cool Dudes: The Denial of Climate Change among Conservative White Males in the United States," *Global Environmental Change* 21/4 (2011): 1163–72.

17 Victor B. Flatt, "Breaking the Vicious Circle: A Review," *Environmental Law* 24/4 (1994): 1707–28.
18 Darrin R. Lehman and Shelley E. Taylor, "Date with an Earthquake: Coping with a Probable, Unpredictable Disaster," *Personality and Social Psychology Bulletin* 13/4 (1987): 546–55.
19 Mary O'Brien, *Making Better Environmental Decisions* (Cambridge, MA: MIT Press, 2000).
20 Ronald W. Rogers and C. Ronald Mewborn, "Fear Appeals and Attitude Change: Effects of a Threat's Noxiousness, Probability of Occurrence, and the Efficacy of Coping Responses," *Journal of Personality and Social Psychology* 34/1 (1976): 54–61.
21 Peter M. Haas, "Introduction: Epistemic Communities and International Policy Coordination," *International Organization* 46/1 (1992): 1–35.
22 Sheila Jasanoff, *The Fifth Branch: Science Advisors as Policymakers* (Cambridge, MA: Harvard University Press, 1990).
23 Nathan Caplan, "The Two-Communities Theory and Knowledge Utilization," *American Behavioral Scientist* 22/3 (1979): 459–70.
24 Daniel F. Zaleznik, *Four Traits of Strong Science–Policy Interfaces for Global Environmental Governance*, MS thesis, University of Massachusetts, 2014.
25 Doyle Rice, "The Last Time the Earth Was This Warm Was 125,000 Years Ago," *USA Today*, 18 January 2017, www. usatoday.com/story/weather/2017/01/18/hottest-year-on-record/96713338/; Daniel Engber, "FYI: What's the Hottest the Earth's Ever Gotten?" *Popular Science*, 5 July 2012, www.popsci. com/science/article/2012-06/fyi-what%E2%80%99s-hottest-earth-has-ever-gotten.
26 Dale Jamieson, "Scientific Uncertainty and the Political Process," *Annals of the American Academy of Political and Social Sciences* 545 (1996): 35–43.
27 Georgina Gustin, "Summer Nights Are Getting Hotter: Here's Why That's a Health and Wildfire Risk," *Inside Climate News*, 11 July 2018, https://insideclimatenews.org/news/09072018/heat-waves-global-warming-overnight-high-temperatures-impact-health-wildlife-wildfires-agriculture.
28 David A. Vasseur, John P. DeLong, Benjamin Gilbert, Hamish S. Greig, Christopher D. G. Harley, Kevin S. McCann, Van Savage,

Tyler D. Tunney, and Mary I. O'Connor, "Increased Temperature Variation Poses a Greater Risk to Species than Climate Warming," *Proceedings of the Royal Society of London B: Biological Sciences* 281 (2014), https://royalsocietypublishing.org/doi/full/10.1098/rspb.2013.2612.

29 John Cook, "Did Global Warming Stop in ..." *Skeptical Science*, 2019, www.skepticalscience.com/global-cooling-january-2007-to-january-2008.htm.

30 Paula A. Johnson, Therese Fitzgerald, Alina Salganicoff, Susan F. Wood, and Jill M. Goldstein, *Sex-Specific Medical Research: Why Women's Health Can't Wait* (Boston: Brigham and Women's Hospital, 2014), https://givingcompass.org/wp-content/uploads/2017/08/ConnorsReportFINAL.pdf.

31 Brent Sparling, "Ozone Depletion, History and Politics," *NASA*, 2001, www.nas.nasa.gov/About/Education/Ozone/history.html.

32 Jamieson, "Scientific Uncertainty and the Political Process."

33 Bjørn Lomborg, *The Skeptical Environmentalist: Measuring The Real State of the World* (Cambridge: Cambridge University Press, 2003).

34 Nebojša Nakićenović and Robert Swart, eds, *Special Report on Emissions Scenarios: A Special Report of Working Group III of the Intergovernmental Panel on Climate Change* (Cambridge: Cambridge University Press, 2000).

35 J. Samuel Barkin and Elizabeth R. DeSombre, *Saving Global Fisheries: Reducing Fishing Capacity to Promote Sustainability* (Cambridge, MA: MIT Press, 2013).

36 Stefanie Tye and Juan-Carlos Altamirano, "Embracing the Unknown: Understanding Climate Change Uncertainty," World Resources Institute, 30 March 2017, www.wri.org/blog/2017/03/embracing-unknown-understanding-climate-change-uncertainty.

37 Riley E. Dunlap and Peter J. Jacques, "Climate Change Denial Books and Conservative Think Tanks: Exploring the Connection," *American Behavioral Scientist* 57/6 (2013): 699–731; Peter J. Jacques, Riley E. Dunlap, and Mark Freeman, "The Organisation of Denial: Conservative Think Tanks and Environmental Scepticism," *Environmental Politics* 17/3 (2008): 349–85.

38 Tristram Korten, "In Florida, Officials Ban Term 'Climate Change,'" *Miami Herald*, 8 March 2015, www.miamiherald.com/news/state/florida/article12983720.html.

39 Alon Harish, "New Law in North Carolina Bans Latest Scientific Predictions of Sea-Level Rise," *ABC News*, 2 August 2012, https://abcnews.go.com/US/north-carolina-bans-latest-science-rising-sea-level/story?id=16913782.
40 Bill McKibben, "The Trump Administration's Solution to Climate Change: Ban the Term," *The Guardian*, 8 August 2017, www.theguardian.com/commentisfree/2017/aug/08/trump-administration-climate-change-ban-usda.
41 Neil deGrasse Tyson, "Earth Needs a Virtual Country: #Rationalia, with a One-Line Constitution: All Policy Shall be Based on the Weight of Evidence," 29 June 2016, https://twitter.com/neiltyson/status/748157273789300736?lang=en.
42 Susan Owens, "Making a Difference? Some Perspectives on Environmental Research and Policy," *Transactions of the Institute of British Geographers* 30/3 (2005): 287–92.

Chapter 3 Political Structures

1 Arthur P. J. Mol, "Sustainability as Global Attractor: The Greening of the 2008 Beijing Olympics," *Global Networks* 10/4 (2010): 510–28.
2 Russell J. Dalton, Doh C. Sin, and Willy Jou, "Understanding Democracy: Data from Unlikely Places," *Journal of Democracy* 18/4 (2007): 142–56.
3 Freedom House, "Freedom in the World 2018: Methodology," https://freedomhouse.org/report/methodology-freedom-world-2018.
4 Laza Kekik, "The Economist Intelligence Unit's Index of Democracy," *The World in 2007*, www.economist.com/media/pdf/DEMOCRACY_INDEX_2007_v3.pdf.
5 Center for Systemic Peace, "The Polity Project," 2016, www.systemicpeace.org/polityproject.html.
6 Juan José Linz, *Totalitarian and Authoritarian Regimes* (Boulder, CO: Lynne Rienner, 2000).
7 Demosthenes James Peterson, *Troubled Lands: The Legacy of Soviet Environmental Destruction* (Boulder, CO: Westview Press, 1993).
8 Paul Robert Josephson, *Resources under Regimes: Technology, Environment, and the State* (Cambridge, MA: Harvard University Press, 2004).

9 Ibid.

10 Timothy Doyle and Adam Simpson, "Traversing More than Speed Bumps: Green Politics under Authoritarian Regimes in Burma and Iran," *Environmental Politics* 15/5 (2006): 750–67.

11 Jane I. Dawson, "Anti-Nuclear Activism in the USSR and its Successor States: A Surrogate for Nationalism?" *Environmental Politics* 4/3 (1995): 441–66.

12 Bruce Gilley, "Authoritarian Environmentalism and China's Response to Climate Change," *Environmental Politics* 21/2 (2012): 287–307, at p. 288.

13 Robert L. Heilbroner, *An Inquiry into the Human Prospect* (London: Calder & Boyers, 1974).

14 Mark Beeson, "Coming to Terms with the Authoritarian Alternative: The Implications and Motivations of China's Environmental Policies," *Asia & The Pacific Policy Studies* 5(1) (2018): 34-46.

15 Freedom House, "Freedom in the World 2019 – Singapore," https://freedomhouse.org/report/freedom-world/2019/singapore.

16 Heejin Han, "Singapore, a Garden City: Authoritarian Environmentalism in a Developmental State," *Journal of Environment & Development* 26/1 (2017): 3–24.

17 Judith Shapiro, *China's Environmental Challenges* (2nd edn, Cambridge: Polity, 2016).

18 Per G. Fredriksson, Eric Neumayer, Richard Damania, and Scott Gates, "Environmentalism, Democracy, and Pollution Control," *Journal of Environmental Economics and Management* 49/2 (2005): 343–65.

19 Madhusudan Bhattarai and Michael Hammig, "Institutions and the Environmental Kuznets Curve for Deforestation: A Crosscountry Analysis for Latin America, Africa and Asia," *World Development* 29/6 (2001): 995–1010.

20 Meilanie Buitenzorgy and Arthur P. J. Mol, "Does Democracy Lead to a Better Environment? Deforestation and the Democratic Transition Peak," *Environmental and Resource Economics* 48/1 (2011): 59–70.

21 Partha Dasgupta and Karl-Göran Mäler, "Poverty, Institutions, and the Environmental Resource-Base," in J. Behrman and T. N. Srinivasan, eds, *Handbook of Development Economics*, vol. 3A (Amsterdam: Elsevier Science, 1995): 2371–463.

22 Kevin P. Gallagher and Strom C. Thacker, *Democracy, Income, and Environmental Quality*, PERI Working Paper, no. 164 (2008), https://scholarworks.umass.edu/cgi/viewcontent. cgi?article=1135&context=peri_workingpapers.

23 Buitenzorgy and Mol, "Does Democracy Lead to a Better Environment?"

24 Eric Neumayer, "Do Democracies Exhibit Stronger International Environmental Commitment? A Cross-Country Analysis," *Journal of Peace Research* 39/2 (2002): 139–64.

25 Roger D. Congleton, "Political Institutions and Pollution Control," *Review of Economics and Statistics* 74/3 (1992): 412–21.

26 Manus I. Midlarsky, "Democracy and the Environment: An Empirical Assessment," *Journal of Peace Research* 35/3 (1998): 341–61, among others.

27 Buitenzorgy and Mol, "Does Democracy Lead to a Better Environment?"

28 Jacqueline M. Klopp, "Deforestation and Democratization: Patronage, Politics and Forests in Kenya," *Journal of Eastern African Studies* 6/2 (2012): 351–70.

29 David Carruthers, "Environmental Politics in Chile: Legacies of Dictatorship and Democracy," *Third World Quarterly* 22/3 (2001): 343–58.

30 Miranda A. Schreurs, "Democratic Transition and Environmental Civil Society: Japan and South Korea Compared," *The Good Society* 11/2 (2002): 57–64.

31 Richard York, "De-Carbonization in Former Soviet Republics, 1992–2000: The Ecological Consequences of De-Modernization," *Social Problems* 55/3 (2008): 370–90.

32 World Bank, *World Development Report 1997: The State in a Changing World* (Oxford: Oxford University Press, 1997).

33 Transparency International, "Corruption Perceptions Index 2018," www.transparency.org/cpi2018.

34 Neal D. Woods, "The Policy Consequences of Political Corruption: Evidence from State Environmental Programs," *Social Science Quarterly* 89/1 (2008): 258–71.

35 Lorenzo Pellegrini, *Corruption, Development and the Environment* (Berlin: Springer Science & Business Media, 2011).

36 Ramon Lopez and Siddhartha Mitra, "Corruption, Pollution,

and the Kuznets Environment Curve," *Journal of Environmental Economics and Management* 40/2 (2000): 137–50.

37 Per G. Fredriksson, Eric Neumayer, and Gergely Ujhelyi, "Kyoto Protocol Cooperation: Does Government Corruption Facilitate Environmental Lobbying?" *Public Choice* 133/1–2 (2007): 231–51.

38 Richard Damania, Per G. Fredriksson, and John A. List, "Trade Liberalization, Corruption, and Environmental Policy Formation: Theory and Evidence," *Journal of Environmental Economics and Management* 46/3 (2003): 490–512.

39 OECD, "Income Inequality," n.d., https://data.oecd.org/inequality/income-inequality.htm; Index Mundi, "Gini Index (World Bank Estimate) – Country Ranking," n.d., www.indexmundi.com/facts/indicators/SI.POV.GINI/rankings.

40 See, among others, Nico Heerink, Abay Mulatu, and Erwin Bulte, "Income Inequality and the Environment: Aggregation Bias in Environmental Kuznets Curves," *Ecological Economics* 38/3 (2001): 359–67.

41 Elisabetta Magnani, "The Environmental Kuznets Curve, Environmental Protection Policy and Income Distribution," *Ecological Economics* 32/3 (2000): 431–43.

42 Matthieu Clement and Andre Meunie, "Is Inequality Harmful for the Environment? An Empirical Analysis Applied to Developing and Transition Countries," *Review of Social Economy* 68/4 (2010): 413–45.

43 John O'Neill, "Wilderness, Cultivation and Appropriation," *Philosophy & Geography* 5/1 (2002): 35–50; Lynn Meskell, *The Nature of Heritage: The New South Africa* (Hoboken, NJ: John Wiley, 2011).

44 Prakash Kashwan, "Inequality, Democracy, and the Environment: A Cross-National Analysis," *Ecological Economics* 131 (2017): 139–51.

45 Thomas O. Wiedmann, Heinz Schandl, Manfred Lenzen, Daniel Moran, Sangwon Suh, James West, and Keiichiro Kanemoto, "The Material Footprint of Nations," *Proceedings of the National Academy of Sciences* 112/20 (2015): 6271–6.

46 Kerry Ard, "Trends in Exposure to Industrial Air Toxins for Different Racial and Socioeconomic Groups: A Spatial and Temporal Examination of Environmental Inequality in the US from 1995 to 2004," *Social Science Research* 53 (2015): 375–90.

47 Rozelia S. Park, "An Examination of International Environmental Racism through the Lens of Transboundary Movement of Hazardous Wastes," *Indiana Journal of Global Legal Studies* 5/2 (1997): 659–709.

48 Ta-Nehesi Coates, "The Case for Reparations," *The Atlantic*, June 2014, www.theatlantic.com/magazine/archive/2014/06/the-case-for-reparations/361631/.

49 Evan J. Ringquist, "Environmental Justice: Normative Concerns, Empirical Evidence, and Government Action," in Norman J. Vig and Michael E. Kraft, *Environmental Policy: New Directions for the Twenty-First Century* (Washington, DC: CQ Press, 2006), pp. 239–52.

50 David M. Konisky, "Regulatory Competition and Environmental Enforcement: Is There a Race to the Bottom?" *American Journal of Political Science* 51/4 (2007): 853–72.

51 Richard L. Revesz, "The Race to the Bottom and Federal Environmental Regulation: A Response to Critics," *Minnesota Law Review* 82/2 (1997): 535–654.

52 David Vogel, *Trading Up: Consumer and Environmental Regulation in a Global Economy* (Cambridge, MA: Harvard University Press, 2009).

53 Matthew Potoski, "Clean Air Federalism: Do States Race to the Bottom?" *Public Administration Review* 61/3 (2001): 335–43.

54 David Vogel, "Representing Diffuse Interests in Environmental Policymaking," in R. Kent Weaver and Bert A. Rockman, eds, *Do Institutions Matter? Government Capabilities in the United States and Abroad* (Washington, DC: Brookings Institution, 1993), pp. 237–71.

55 Detlef Jahn and Ferdinand Müller-Rommel, "Political Institutions and Policy Performance: A Comparative Analysis of Central and Eastern Europe," *Journal of Public Policy* 30/1 (2010): 23–44.

56 Gary Cox and Mathew McCubbins, "Political Structure and Economic Policy: The Institutional Determinants of Policy Outcomes," in Stephan Haggard and Matthew D. McCubbins, eds, *Presidents, Parliaments, and Policy* (Cambridge: Cambridge University Press, 2001), pp. 21–63.

57 Olle Folke, "Shades of Brown and Green: Party Effects in Proportional Election Systems," *Journal of the European Economic Association* 12/5 (2014): 1361–95.

58 Per G. Fredriksson and Daniel L. Millimet, "Electoral Rules and Environmental Policy," *Economics Letters* 84/2 (2004): 237–44.

59 Robert Rohrschneider, "New Party versus Old Left Realignments: Environmental Attitudes, Party Policies, and Partisan Affiliations in Four West European Countries," *Journal of Politics* 55/3 (1993): 682–701.

60 Theda Skocpol, "Bringing the State Back In: Strategies of Analysis in Current Research," in P. Evans, D. Rueschemeyer, and T. Skocpol, eds, *Bringing the State Back In* (New York: Cambridge University Press, 1990), pp. 3–43.

61 Hanna Bäck and Axel Hadenius, "Democracy and State Capacity: Exploring a J-Shaped Relationship," *Governance* 21/1 (2008): 1–24.

62 George Tsebelis, "Decision Making in Political Systems: Veto Players in Presidentialism, Parliamentarism, Multicameralism and Multipartyism," *British Journal of Political Science* 25/3 (1995): 289–325.

63 Nathan J. Madden, "Green Means Stop: Veto Players and Their Impact on Climate-Change Policy Outputs," *Environmental Politics* 23/4 (2014): 570–89.

CHAPTER 4 POLITICAL ACTORS

1 Riley E. Dunlap and Aaron M. McCright, "Social Movement Identity: Validating a Measure of Identification with the Environmental Movement," *Social Science Quarterly* 89/5 (2008): 1045–65.

2 Jennifer E. Givens and Andrew K. Jorgenson, "Individual Environmental Concern in the World Polity: A Multilevel Analysis," *Social Science Research* 42/2 (2013): 418–31.

3 Ion Bogdan Vasi, "New Heroes, Old Theories? Toward a Sociological Perspective on Social Entrepreneurship," in R. Ziegler, ed., *An Introduction to Social Entrepreneurship* (Cheltenham: Edward Elgar, 2009), pp. 155–73.

4 B. Dan Wood, "Principals, Bureaucrats, and Responsiveness in Clean Air Enforcements," *American Political Science Review* 82/1 (1988): 213–34.

5 Jon Owens, "Comparative Law and Standing to Sue: A Petition for Redress for the Environment," *Environmental Law* 7/2 (2000): 321–77.
6 Rob White, "Reparative Justice, Environmental Crime and Penalties for the Powerful," *Crime, Law and Social Change* 67/2 (2017): 117–32.
7 *Massachusetts* v. *Environmental Protection Agency*, 549 US 497 (2007).
8 Jonathan Stempel, "U.S. States Sue EPA, Pruitt for Rolling Back Climate Change Rule," *Reuters*, 27 June 2018, www.reuters.com/article/us-usa-climatechange-lawsuit/u-s-states-sue-epa-pruitt-for-rolling-back-climate-change-rule-idUSKBN1JN2UO.
9 Lois J. Schiffer and Timothy J. Dowling, "Reflections on the Role of the Courts in Environmental Law," *Environmental Law* 27 (1997): 327–42.
10 Patricia M. Wald, "The Role of the Judiciary in Environmental Protection," *Boston College Environmental Affairs Law Review* 19/3 (1991): 519–46.
11 Karen Litfin, "Ecoregimes: Playing Tug of War with the Nation-State," in Ronnie Lipschutz and Ken Conca, eds, *The State and Social Power in Global Environmental Politics* (New York: Columbia University Press, 1993), p. 100.
12 Paul Wapner, *Environmental Activism and World Civic Politics* (Albany: SUNY Press, 1996).
13 Elizabeth R. DeSombre, "Distorting Global Governance: Membership, Voting, and the IWC," in Robert Friedheim, ed., *Toward a Sustainable Whaling Regime* (Seattle: University of Washington Press, 2001), pp. 183–99.
14 Pamela S. Chasek, *Earth Negotiations: Analyzing Thirty Years of Environmental Diplomacy* (New York: United Nations University Press, 2001).
15 See, for example, Alex Chadwick, "The Treasured Islands of Palmyra," *National Geographic*, March 2001, pp. 46–56.
16 Cord Jacobeit, "Nonstate Actors Leading the Way: Debt-for-Nature Swaps," in Robert O. Keohane and Marc A. Levy, eds, *Institutions for Environmental Aid* (Cambridge, MA: MIT Press, 1997), pp. 127–66.
17 Russel Hardin, *Collective Action* (Baltimore: Johns Hopkins University Press, 1982).

18 Charles Edward Lindblom, *The Policy-Making Process* (Upper Saddle River, NJ: Prentice-Hall, 1968).

19 Michael E. Porter, "America's Green Strategy," *Scientific American*, 264/4 (1991): 168.

20 Livio D. DeSimone and Frank Popoff, *Eco-Efficiency: The Business Link to Sustainable Development* (Cambridge, MA: MIT Press, 1997), pp. 31, 39–40.

21 Frances Cairncross, "Cleaning Up," *The Economist*, 8 September 1990, pp. S1ff.

22 Quoted in Kenny Bruno, "The Corporate Capture of the Earth Summit," *Multinational Monitor* 13 (July/August 1992), p. 18.

23 Michael S. Baram, "Multinational Corporations, Private Codes, and Technology Transfer for Sustainable Development," *Environmental Law* 24/1 (1994): 33–66.

24 Elizabeth R. DeSombre, *Domestic Sources of International Environmental Policy: Industry, Environmentalists and U.S. Power* (Cambridge, MA: MIT Press, 2000).

25 Peter J. Jacques, Riley E. Dunlap, and Mark Freeman, "The Organisation of Denial: Conservative Think Tanks and Environmental Scepticism," *Environmental Politics* 17/3 (2008): 349–85.

26 Naomi Oreskes, "The Scientific Consensus on Climate Change," *Science* 306 (2004): 1686.

27 Maxwell T. Boykoff and Jules M. Boykoff, "Balance as Bias: Global Warming and the US Prestige Press," *Global Environmental Change* 14/2 (2004): 125–36.

CHAPTER 5 INTERNATIONAL ENVIRONMENTAL POLITICS

1 Robert D. Putnam, "Diplomacy and Domestic Politics: The Logic of Two-Level Games," *International Organization* 42/3 (1988): 427–60.

2 US Congress, Senate, 105th Congress, 1st session, *Congressional Record*, daily edn, 27 July 1997, S8113–8138.

3 Joe Deverian, "President Obama Bypasses Congress by Formally Committing U.S. to Paris Climate Agreement," *American*

Legislative Exchange Council, 22 September 2016, www.alec.org/
article/president-obama-bypasses-congress-by-formally-committing-
u-s-to-paris-climate-agreement/.

4 Bill Curry and Shawn McCarthy, "Canada Formally Abandons
Kyoto Protocol on Climate Change," *Globe and Mail*, 12 December
2011, www.theglobeandmail.com/news/politics/canada-formally-
abandons-kyoto-protocol-on-climate-change/article4180809/.

5 Euan McKirdy, Emiko Jozuka, and Junko Ogura, "IWC
Withdrawal: Japan to Resume Commercial Whaling in 2019,"
CNN, 26 December 2018, https://edition.cnn.com/2018/12/25/asia/
japan-withdrawal-international-whaling-commission-intl/index.
html.

6 Chris Mooney, "Trump Can't Actually Exit the Paris Deal
Until the Day After the 2020 Election: That's a Big Deal,"
Washington Post, 12 December 2018, www.washingtonpost.com/
energy-environment/2018/12/12/heres-what-election-means-us-
withdrawal-paris-climate-deal.

7 Roger Fisher, *Improving Compliance with International Law*
(Charlottesville: University of Virginia Press, 1981).

8 Louis Henkin, *How Nations Behave: Law and Foreign Policy* (New
York: Columbia University Press, 1979).

9 Rosalind Reeve, *Policing International Trade in Endangered
Species: The CITES Treaty and Compliance* (London: Routledge,
2014).

10 Virginia M. Walsh, "Illegal Whaling for Humpbacks by the Soviet
Union in the Antarctic, 1947–1972," *Journal of Environment &
Development* 8/3 (1999): 307–27.

11 Eric Neumayer, "How Regime Theory and the Economic Theory
of International Environmental Cooperation Can Learn From Each
Other," *Global Environmental Politics* 1/1 (2001): 122–47.

12 Edward A. Parson, *Protecting the Ozone Layer: Science and
Strategy* (Oxford: Oxford University Press, 2003).

13 Elizabeth R. DeSombre and Joanne Kauffman, "The Montreal
Protocol Multilateral Fund: Partial Success Story," in Robert O.
Keohane and Marc A. Levy, eds, *Institutions for Environmental
Aid* (Cambridge, MA: MIT Press, 1996), pp. 89–126.

14 Ozone Secretariat, "Montreal Protocol – Achievements to Date
and Challenges Ahead," 2015, http://42functions.net/en/MP_
achievements_challenges.php.

15 Environment and Climate Change Canada, "Ozone-Depleting Substances," www.ec.gc.ca/ozone/default.asp?lang=En&n=D57A0006.

16 World Meteorological Organization, *Scientific Assessment of Ozone Depletion: 2018*, Global Ozone Research and Monitoring Project – report no. 58 (Geneva: WMO, 2018).

17 Elizabeth R. DeSombre, "The Experience of the Montreal Protocol: Particularly Remarkable and Remarkably Particular," *UCLA Journal of Environmental Law and Policy*, 19/2 (2001): 49–81.

18 Chris Buckley, "More Evidence Points to China as Source of Ozone-Depleting Gas," *New York Times*, 3 November 2018, www.nytimes.com/2018/11/03/climate/china-ozone-cfcs.html.

19 Radoslav S. Dimitrov, "Hostage to Norms: States, Institutions, and Global Forest Politics," *Global Environmental Politics* 5/4 (2005): 1–24.

20 Stacy D. VanDeveer, "Green Fatigue," *Wilson Quarterly* 27/4 (2003): 55–9.

21 Thijs Van de Graaf, "Fragmentation in Global Energy Governance: Explaining the Creation of IRENA," *Global Environmental Politics* 13/3 (2013): 14–33.

22 Alexander Gillespie, "Forum Shopping in International Environmental Law: The IWC, CITES, and the Management of Cetaceans," *Ocean Development & International Law* 33/1 (2002): 17–56.

23 William Alexander Kerr, Stuart J. Smyth, Peter W. B. Phillips, and Martin Phillipson, "Conflicting Rules for the International Trade of GM Products: Does International Law Provide a Solution?" *AgBioForum* 17/2 (2014): 105–22.

24 Rodney Bruce Hall and Thomas J. Biersteker, *The Emergence of Private Authority in Global Governance* (Cambridge: Cambridge University Press, 2002).

25 JoAnne Yates and Craig N. Murphy, *Engineering Rules: Global Standard Setting since 1880* (Baltimore: Johns Hopkins University Press, 2019).

26 Graeme Auld, *Constructing Private Governance: The Rise and Evolution of Forest, Coffee, and Fisheries Certification* (New Haven, CT: Yale University Press, 2014).

27 FSC, "FSC Facts and Figures," 3 January 2019, https://ic.fsc.org/en/facts-and-figures.

28 ICLEI, "Our Activities," https://iclei.org/en/featured_activities. html.
29 Jessica F. Green, *Rethinking Private Authority: Agents and Entrepreneurs in Global Environmental Governance* (Princeton, NJ: Princeton University Press, 2013).
30 Clean Development Mechanism, "List of DOEs," https://cdm. unfccc.int/DOE/list/index.html.
31 TRAFFIC, "About Us – Working with CITES," www.traffic.org/ about-us/working-with-cites/.
32 Steve Maguire and Cynthia Hardy, "The Emergence of New Global Institutions: A Discursive Perspective," *Organization Studies* 27/1 (2006): 7–29.
33 Elizabeth R. DeSombre, *Domestic Sources of International Environmental Policy: Industry, Environment, and U.S. Power* (Cambridge, MA: MIT Press, 2000).

CHAPTER 6 ENGAGING WITH ENVIRONMENTAL POLITICS

1 Winston Churchill speaking in the House of Commons, 11 November 1947.
2 United Nations Environment Programme, *GEO 6: Healthy Planet, Healthy People* (Cambridge: Cambridge University Press, 2019).
3 United Nations Food and Agriculture Organization, *State of World Fisheries and Aquaculture 2018: Meeting the Sustainable Development Goals* (Rome: FAO, 2018).
4 Ransom A. Myers and Boris Worm, "Rapid Worldwide Depletion of Predatory Fish Communities," *Nature* 423 (2003): 280–3.
5 Lijing Cheng, John Abraham, Zeke Hausfather, and Kevin E. Trenberth, "How Fast are the Oceans Warming?" *Science* 363 (2019): 128–9.
6 Intergovernmental Panel on Climate Change, *Global Warming of 1.5°C* (Geneva: IPCC, 2018).
7 Global Burden of Disease Collaborative Network, *Global Burden of Disease Study 2016 (GBD 2016) Results* (Seattle: Institute for Health Metrics and Evaluation (IHME), 2017).

8 United States Environmental Protection Agency, "Air Emissions Inventory – Air Pollutant Emissions Trends Data," 2017, www.epa. gov/air-emissions-inventories/air-pollutant-emissions-trends-data.

9 United Nations, *Millennium Development Goals Report 2015* (New York: United Nations, 2015), p. 52.

10 Steven Pinker, *Enlightenment Now: The Case for Reason, Science, Humanism, and Progress* (New York: Penguin Books, 2018).

11 Johnny Briggs, Stacy K. Baez, Terry Dawson, Bronwen Golder, Bethan C. O'Leary, Jerome Petit, Callum M. Roberts, Alex Rogers, and Angelo Villagomez, "Recommendations to IUCN to Improve Marine Protected Area Classification and Reporting," 6 February 2018, Pew Bertarelli Ocean Legacy Project, www.pewtrusts.org/-/ media/assets/2018/02/recommendations-to-iucn-on-implementing-mpa-categories-for-printing.pdf.

12 United Nations, *Millennium Development Goals Report 2015*.

13 United Nations Economic Commission for Europe, "40 Years of Successful Cooperation for Clean Air," 2019, www.unece.org/ environmental-policy/conventions/envlrtapwelcome/40-years-clean-air.html.

14 ITOPF, "Oil Tanker Spill Statistics 2018," www.itopf.org/ knowledge-resources/documents-guides/document/oil-tanker-spill-statistics-2018/.

15 Axel Franzen and Dominikus Vogl, "Two Decades of Measuring Environmental Attitudes: A Comparative Analysis of 33 Countries," *Global Environmental Change* 23/5 (2013): 1001–8.

16 Lorraine Whitmarsh and Saffron O'Neill, "Green Identity, Green Living? The Role of Pro-Environmental Self-Identity in Determining Consistency across Diverse Pro-Environmental Behaviours," *Journal of Environmental Psychology* 30/3 (2010): 305–14.

17 Elizabeth R. DeSombre, *Why Good People Do Bad Environmental Things* (Oxford: Oxford University Press, 2018).

18 Matthew Feinberg and Robb Willer, "Apocalypse Soon? Dire Messages Reduce Belief in Global Warming by Contradicting Just-World Beliefs," *Psychological Science* 22/1 (2011): 34–8.

19 Daniel J. Fiorino, *Can Democracy Handle Climate Change?* (Cambridge: Polity, 2018).

20 David Daley, "Maine's 2nd District Outcome Proves Value of Ranked Choice Voting," *The Hill*, 30 November 2018, https://thehill.com/opinion/

campaign/418058-maines-2nd-district-outcome-proves-value-of-ranked-choice-voting.

21 Elizabeth R. DeSombre, *Domestic Sources of International Environmental Policy: Industry, Environmentalists, and U.S. Power* (Cambridge, MA: MIT Press, 2000).

Selected Readings

This brief book has only scratched the surface of environmental politics; there are many excellent sources to consult for a deeper look. To think about *why* we need to be addressing environmental issues, some excellent options include David Wallace-Wells, *The Uninhabitable Earth: Life after Warming* (New York: Tim Duggan Books, 2019), and Elizabeth Kolbert, *The Sixth Extinction: An Unnatural History* (New York: Henry Holt, 2014). And on why individual action will not suffice to address environmental problems, check out Elizabeth R. DeSombre, *Why Good People Do Bad Environmental Things* (Oxford: Oxford University Press, 2018).

On the framing of environmental issues as externalities and as collective action and common pool resource issues, a classic of the field is Elinor Ostrom, *Governing the Commons: The Evolution of Institutions for Collective Action* (Cambridge: Cambridge University Press, 1990). Another excellent source (which focuses on international issues but is also more broadly relevant) is J. Samuel Barkin and George E. Shambaugh, eds, *Anarchy and the Environment: The International Relations of Common Pool Resources* (Albany: SUNY Press, 1999). For more information about the question of whether we will run out of resources (and the Ehrlich–Simon wager more generally), the best overview is Paul Sabin, *The Bet: Paul Ehrlich, Julian Simon, and Our Gamble over Earth's Future* (New Haven, CT: Yale University Press, 2013).

The relationship between science and environmental politics has also been widely and capably studied. Among the most important early works are Kai N. Lee, *Compass and Gyroscope: Integrating Science and Politics for the Environment* (Washington, DC: Island Press, 1993), and Sheila Jasanoff, *The Fifth Branch: Science Advisers as Policymakers* (Cambridge, MA: Harvard University Press, 2009). Peter Haas has been instrumental in conceptualizing the role scientists can play in international environmental politics; in *Saving the Mediterranean: the Politics of International Environmental Cooperation* (New York: Columbia University Press, 1990), he developed the idea of epistemic communities. Finally, the case of climate change denial is a worrisome and important topic that is beginning to get more attention. Among the first books on this topic, framing it more broadly about environmental skepticism, is Peter Jacques, *Environmental Skepticism: Ecology, Power and Public Life* (London: Routledge, 2009). Another early classic that addresses how doubt is intentionally manufactured across a number of issues is Naomi Oreskes and Erik M. Conway, *Merchants of Doubt: How a Handful of Scientists Obscured the Truth on Issues from Tobacco Smoke to Global Warming* (London: Bloomsbury, 2011).

An excellent starting point for the politics of the environment in comparative perspective is Paul F. Steinberg and Stacy D. VanDeveer, eds, *Comparative Environmental Politics: Theory, Practice, and Prospects* (Cambridge, MA: MIT Press, 2012). And a thorough overview of the conceptual and empirical relationship between democracy and environmental protection can be found in William M. Lafferty and James Meadowcroft, *Democracy and the Environment* (Cheltenham: Edward Elgar, 1996). For imperfections in democracy, there is, thankfully, growing scholarly attention to the issue of environmental (in) justice, including the excellent and quite recent overview by

David Naguib Pellow, *What is Critical Environmental Justice?* (Cambridge: Polity, 2018), as well as Steve Vanderheiden, *Environmental Justice* (London: Routledge, 2017), and the now-classic Robert D. Bullard, *Dumping in Dixie: Race, Class, and Environmental Quality* (Boulder, CO: Westview Press, 2008).

This volume has only touched on some of the important institutions and political actors and their relationship to environmental politics. Some sources that take these introductions further are James Connelly, Graham Smith, David Benson, and Clare Saunders, *Politics and the Environment: from Theory to Practice* (3rd edn, London: Routledge, 2012), and Timothy Doyle, Doug McEachern, and Sherilyn MacGregor, *Environment and Politics* (London: Routledge, 2015). Also useful for further understanding the role of business is Peter Dauvergne, *Will Big Business Destroy Our Planet?* (Cambridge: Polity, 2018). One of the most insightful scholars these days on the role of activists and non-governmental organizations in politics (on environmental and other issues) is Hahrie Han. Her book *How Organizations Develop Activists: Civic Associations and Leadership in the 21st Century* (New York: Oxford University Press, 2014) and also *Moved to Action: Motivation, Participation, and Inequality in American Politics* (Stanford, CA: Stanford University Press, 2009), though both are focused on the United States, are worth consulting whatever geographic area you study. The interaction between business and environmentalists is ably covered in Jem Bendell, *Terms for Endearment: Business, NGOs and Sustainable Development* (London: Routledge, 2017).

Finally, there are some books that usefully take the framing of international environmental politics beyond the state-centric focus that predominates (including here). Simon Nicholson and Paul Wapner, eds, *Global Environmental*

Politics: From Person to Planet (London: Routledge, 2015), is an excellent set of essays that explain why we need to care about global environmental politics. Two other books that cover the broad framing of the intersection between international political economy and the politics of the global environment from multiple perspectives are Jennifer Clapp and Peter Dauvergne, *Paths to a Green World: The Political Economy of the Global Environment*, (2nd edn, Cambridge, MA: MIT Press, 2011), and the regularly updated Ken Conca and Geoffrey D. Dabelko, eds, *Green Planet Blues* (6th edn, Boulder, CO: Westview Press, 2019). And, of the many recent discussions of non-state standards and certification, great sources include Jessica F. Green, *Rethinking Private Authority: Agents and Entrepreneurs in Global Environmental Governance* (Princeton, NJ: Princeton University Press, 2013), Graeme Auld, *Constructing Private Governance: The Rise and Evolution of Forest, Coffee, and Fisheries Certification* (New Haven, CT: Yale University Press, 2014), and Lars H. Gulbrandsen, *Transnational Environmental Governance: The Emergence and Effects of the Certification of Forest and Fisheries* (Cheltenham: Edward Elgar, 2010).

Index